Introduction to VHDL

R.D.M. Hunter
University of Portsmouth, UK

and

T.T. Johnson
Summit Design Inc., USA

CHAPMAN & HALL
London · Glasgow · Weinheim · New York · Tokyo · Melbourne · Madra

Published by Chapman & Hall, 2–6 Boundary Row, London SE1 8HN, UK

Chapman & Hall, 2–6 Boundary Row, London SE1 8HN, UK

Blackie Academic & Professional, Wester Cleddens Road, Bishopbriggs, Glasgow G64 2NZ, UK

Chapman & Hall GmbH, Pappelallee 3, 69469 Weinheim, Germany

Chapman & Hall USA, 115 Fifth Avenue, New York, NY 10003, USA

Chapman & Hall Japan, ITP-Japan, Kyowa Building, 3F, 2-2-1 Hirakawacho, Chiyoda-ku, Tokyo 102, Japan

Chapman & Hall Australia, 102 Dodds Street, South Melbourne, Victoria 3205, Australia

Chapman & Hall India, R. Seshadri, 32 Second Main Road, CIT East, Madras 600 035, India

---

First edition 1996

Printed in Great Britain by Clays Ltd, St Ives plc

ISBN 0 412 73130 4

A catalogue record for this book is available from the British Library

∞ Printed on permanent acid-free text paper, manufactured in accordance with ANSI/NISO Z39.48-1992 and ANSI/NISO Z39.48-1984 (Permanence of Paper).

Contents

Preface

The authors of this textbook perceived a need for a low-cost University 'Primer' which can introduce VHDL to undergraduates and others having a requirement for an awareness of the language, and this text seeks to satisfy that need. A basic, but not extensive, knowledge of digital design is expected of the reader.

The book emerges from a set of VHDL Notes created by one of the authors, Tom Johnson, formerly of the Mentor Graphics Corporation®, and written for the edification of colleagues and customers of the Corporation. The version of these Notes presented in this text has been adapted for the introduction of VHDL to undergraduates studying electronic engineering within the Faculty of Technology at the University of Portsmouth in the U.K.

The Department of Electrical and Electronic Engineering at the University of Portsmouth has had a close relationship with MGC for a number of years and has had the privilege of working with the Company to introduce CAD tools to undergraduates at the earliest stage possible (coincident with their introduction to the market place, and, on occasion, even earlier). The University has also co-operated at the development stage of new MGC products for the design and manufacture of electronic systems.

Learning a foreign language requires continuous exposure to its syntax and the study of a systems hardware description language like VHDL is no different. This volume, therefore, uses the technique of repetition for the purpose of reinforcement and sound familiarity. This leads to another feature of the book in that certain examples and exercises build through the chapters, adding complexity as they move forward. For instance, a design example involving the creation of the threshold detector described in Chapter 3 builds from the simplest level through to the final design, employing packages, as an exercise in Chapter 16. It should be noted, however, that the most probable design approach would largely be carried out in the reverse order.

An aim of this book is to present VHDL in a progressive manner, although it is recognised that this approach must limit flexibility towards the presentation of a fully integrated language. This is especially true in respect of the 'end-of-chapter' exercises, some of which might have been more meaningful had they been able to avail themselves of aspects of the language yet to be introduced.

The technique favoured by MGC of illuminating their instruction sessions with bold, illustrative slides (or viewfoils) has been retained in this text and is aimed at lecturers and instructors wishing to introduce VHDL to their students, as much as to the students themselves. The construction of the book uses Illumination Boxes on most left-hand pages, which essentially summarise the text on the right-hand pages, and these provide a means for the reader to 'fast-track' through the basics of the language. They also provide a convenient 'look-up' method of accessing the principles involved since they use the same headings as the chapters themselves.

A solutions manual to accompany the 'end-of-chapter' exercises is available from the authors.

R D M Hunter

T T Johnson

Chapter 1

Introduction

1.1 The need for change

Human beings are capable of a richness and diversity of language, etymology and semantics which is quite staggering. In the complexities and competitiveness of their relationships people describe, prescribe and produce a bewildering array of superficially different solutions to essentially similar problems. Take the car industry; in order to sell cars a manufacturer must locate the critical functions of steering wheel, brake pedal, clutch and accelerator (the human interfaces) in essentially the same positions for rather obvious reasons. Add-on refinements, not being critical, are subject to designer whim as to location and method of operation. The 'language' employed in the manufacture of a car must, therefore, contain some universal truths. Thereafter, in the pursuit of a competitive edge, it can be flamboyant, esoteric, even obscure. The purpose of the latter is to attract a larger share of the market and to distract the competition. The outcome is a product which, in virtually all of its elements, is superficially unique and non-interchangeable between manufacturers and yet essentially similar in function and purpose.

A process which echos the above is evident in the electronics industry. Because of the need to supply to a mass market many components have become standardised and second-sourced throughout the manufacturing community. Once again, however, these components can be assembled into a product which is required to fulfil identical functions as defined by conditions external to itself, e.g. a television receiver, and yet it will be superficially different and non-interchangeable between manufacturers. Furthermore, because of process variations,a product may also have features which render it non-portable between suppliers to a single customer.

In addition to the above, the introduction of more and more powerful computers and workstations has led to the development of a myriad of correspondingly powerful Computer-Aided-Design tools (CAD). Such tools, designed by independent producers, have not necessarily been convergent in the way they interpret design specifications (owing to a need to include competitive features) even when a self-evidently similar purpose might seem to preclude divergence.

Finally, tools designed to complete the loop between specification and product have also been developed by independent vendors and, while a healthy movement of personnel among the leading suppliers has been observed, this has led to the interpretation of specification detail being dependent on the digital circuit expertise of the design team. The outcome has tended to be that the same specification may lead to hardware which, as in the case of the automobile and the TV set, is more or less efficient according to certain criteria (area, speed, gate count etc.) but is non.portable between suppliers to a single customer.

All of the foregoing presents a problem for traceability in the design process where purchaser of a system has a special reason for wishing to keep control of the development and sourcing of the product such as in the case of VLSI circuits in electronic systems for military purposes. It is this dilemma which led the Department of Defense in the United States of America into the setting up of a project for a vendor-independent method of standardising of structures and documentation in the design and manufacture of Very High Speed Integrated Circuits (VHSIC) and the development of a language to support the process, V(HSIC)HDL or VHDL, Very High Speed Integrated Circuit Hardware Description Language.

In order to ensure that the language should enter into the public domain in a controlled fashion the DoD released the language to the IEEE for evaluation and hardening, a process which resulted in the issue of the IEEE standard version of VHDL (1076-1987) in December 1987. Since that time a number of simulators have appeared on the market for the synthesis and analysis of VHDL of which Mentor Graphics System-1076® comprises a very comprehensive example.

The VHDL Language is subject to re-balloting every five years and therefore, at the time of writing, the first re-ballot has taken place. This has resulted in the issue of the 1993 version of the LRM, ANSI/IEEE Std 1076-1993. The VHDL Analysis and Standardisation Group (VASG) issued a set of 48 Language Change Specifications (LCS), most of which have been implemented (Appendix C), and these are included, where relevant, throughout this text.

Basically, then, a textual language for the determination of the structure, or definition, of devices (models) and their connectivity has been created by means of 'Reserved Words' which are themselves unambiguously defined (see Chapter 3, Section 3.1 andAppendix A). Because VHDL is a rich language the number of reserved words is necessarily large (97). It also has a number of features which are common to other recently developed languages such as ADA. These features are the result of a natural convergence of thinking towards the 'ideal' computer language and although VHDL should not be thought of as a software language it has many similarities with the modern concepts being developed for programming. For example, both ADA and VHDL use the concept of objects, an object in ADA being of two types *constant* or *variable*. VHDL is similar, but with the addition of type *signal*. Type *signal* is necessary in order to deal with concurrency (amongst other things) in the modelling of real systems and may be likened to the physical wire in electronic circuits (see Chapter 6, Section 6.3).

As indicated above there may be compelling reasons why the defence organisation of a government would wish to promote a hardware description language such as VHDL but can it be demonstrated also that civil, or commercial, enterprises and, coterminously, digital circuit designers, should have a vested interest in the success of the language?

Many years ago Gordon Moore, of the Intel Corporation, claimed that "The potential system complexity of digital circuits doubles every two years". This prediction shows no indication of slowing down and his more recent claim that "Design time doubles every thirty-two months" maintains a corresponding validity. Add to this the fact that, as the feature size of devices shrinks towards deep sub.micron (<5 micron), the number of potential gates in a single design is increasing exponentially into the millions.

It is these twin trends which have produced the most compelling reason for the adoption of VHDL as the preferred design methodology. The power of modern tools has had its own effect upon competitiveness and has played a pivotal role in drastically reducing what is called 'time-to-market' (TTM). It can be shown (Figure 1.1 below) that there is a directly-coupled relationship between market growth, market life (or market 'window'), delay in reaching the market, and revenue[1].

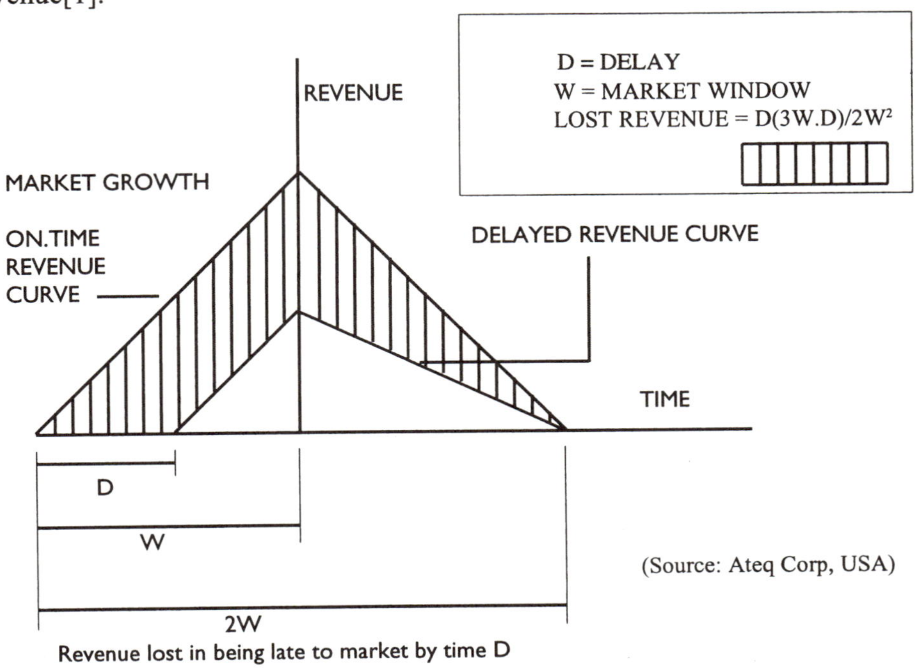

(Source: Ateq Corp, USA)

Figure 1.1 Market Growth vs market life

It is observed that the time available, from conception to product, is reducing rapidly in respect of a prospective vendor having a realistic opportunity of capturing a profitable share of the market for the product. For some electronic products TTM has fallen to less than one year. Furthermore there is no longer any guarantee of a long (>5year) lifetime for the product. In fact, in some product areas, marketable life has also shrunk to under one year. A design survey has revealed that design productivity has been doubling every four years owing to the use of synthesis tools and HDLs, and this trend is set to continue[2]. Even so, productivity is not keeping pace with increasing chip complexity.

How, then, can VHDL help digital circuit designers keep pace with accelerating market demands? To answer this question it is first necessary to reflect upon why the 'traditional' methods of realising a design specification are no longer adequate. Since the late '70s this has usually meant deconstructing a specification down to at least register-transfer level (RTL) and reconstructing towards the final implementation using some form of circuit capture technique such as floor-planning for full-custom design (pre-CAD) or schematic tools at gate level and above, a process described as 'bottom-up' design (See Section 2.1). This has been because, historically, designers were constrained by the availability of vendor-only supplied SSI, MSI, LSI and even VLSI circuit elements. Even with the advent of ASICs schematic capture tools dictated a bottom-up approach to design. Although top-down design strategies were possible the 'black boxes' of such designs were, using this approach, empty shells which merely enabled system connectivity to be established at an early stage. They could only become 'active' by introducing functionality, internally, at the lowest (gate) level.

As designs become more complex, and gate counts increase towards hundreds of thousands, keeping track of the multitude of 'sheets' needed to represent a design is proving to be a near impossible task. It has become clear that the the emerging text-based methods are the only methods available equal to the task of taking advantage of the opportunity presented by the latest techological advances to create large designs within an acceptable time frame.

1.2 The means for change

An HDL offers the opportunity of a true top.down approach to electronic system design in that a design can be simulated or verified, at, or very near to, system level. With the system level tools now available compliance with specification can be checked at the highest level and a direct, unbroken, link constructed down to gate level and sign-off to ASIC vendor or to Programmable Logic Arrays (PLAs). Vendors' device timing constraints are becoming available through the VITAL initiative (See Section 2.2) and back annotation using the Standard Delay Format (SDF) will ensure that models are fully simulatable with respect to device and interconnect propagation delay. All of the above will allow a product to be synthesised such that it can be manufactured with the confidence that it will be 'Right First Time', a prized goal of modern Application Specific Integrated Circuit (ASIC) design.

1.3 Why use VHDL?

VHDL satisfies all the requirements for the hierarchical description of electronic circuits from system level down to switch level. It can support all levels of timing specification and constraints and is capable of detecting and signalling timing violations. The language models the reality of concurrency present in digital systems and supports the recursivity of finite state machines. The concepts of packages and configurations allows the creation of design libraries for the re-use of previously designed parts. A possible structure indicating the scope of VHDL is given in Figure 1.2 below.

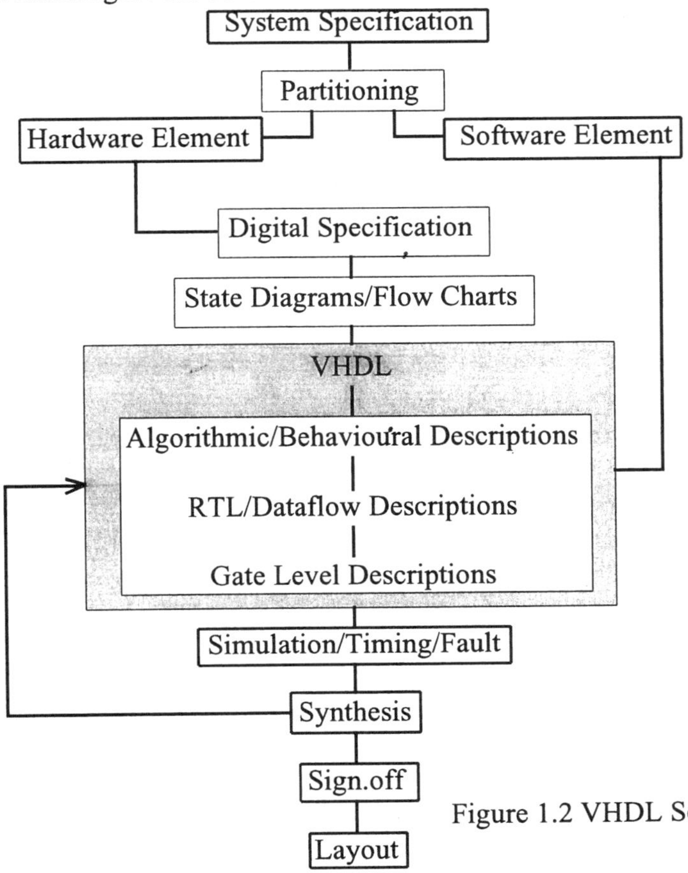

Figure 1.2 VHDL Scope

Used efficiently VHDL will increase productivity by a large factor and TTM will decrease dramatically. Technology independence will allow technology migration to, for example, reduced feature lengths in ASICs (deep sub-micron) or to, say, from a Field Programmable Gate Array (FPGA) to ASIC where an FPGA has been used to prove the functionality of a design. The consequence of reduced time scales will force greater concentration on system issues, not the least of which is the need to improve the integrity, in the form of higher quality and greater precision, of specifications.

It should not be imagined, however, that being able to synthesise digital circuits from VHDL is a panacea for inadequate digital circuit design skills. Efficient design is still the responsibiliy of the designer if he is to obtain the best from the HDL methodology. Synthesis of hardware is very sensitive to the way in which the VHDL code is written and poor descriptions will will give rise to inefficient implementation since synthesisers are not yet sufficiently intelligent to subsume the task of 'second-guessing' the intentions of the designer. As an illustration suppose a designer has created a custom microprocessor in order to eliminate the inefficiency of using a proprietary microprocessor in his application. Poor coding strategies may lead to multiple copies of standard components such as arithmetic logic units being employed. The inability to model the sharing of resources might nullify the advantages to be gained from customising the processor. Better 'handcrafting' of the design could bring about a reduction in the gate count offering the possibility of higher yield in ASIC implementation, lower test costs and potentially increased performance.

1.4 Summary of Chapters

This volume consists of sixteen chapters most of which serve to introduce basic features of VHDL and to develop these through the use of examples and exercises. It attempts to follow the sequence employed in the VHDL Language Reference Manual (LRM)as far as is possible and contains exercises for the reader's practice and some worked examples. A number of the worked examples build through certain chapters as topics are introduced relevant to them. For instance, the design example on the threshold detector begins at the structural level in Chapter 3 and is completed in Chapter 14 where packages are introduced. It should be noted however that the design might actually start with the level described in Chapter 14 and be completed by adding the relevant structural code as described in Chapter 3.

Chapter 2 traces the history of Hardware Description Languages. It explains how schematic capture tools, introduced in the early nineteen eighties, pushed attempts at HDLs into the background until the increasing size of integrated circuits made such methods unwieldy.

This chapter also compares top-down with bottom-up approaches to electronic design and indicates the decisions which have to be made in making the appropriate choice of method to be used for any particular application.

Chapter 3 introduces the basic concepts behind hardware description languages with particular reference to VHDL. The idea of reserved words and identifiers is explained and the context relationships between these are discussed.

Chapter 4 introduces the salient ideas behind VHDL and introduces the concepts of Design Entity, Entity Declaration, Port Statement and Architecture Body.

Chapter 5 deals with objects, data types and operators. These add necessary dimensions and parameters to the language.

Chapter 6 is concerned with the levels of abstraction in VHDL and develops the idea of multiple architectures. This chapter also shows how VHDL handles the reality of concurrent operation in hardware

Chapter 7 concerns itself with sequential operations and introduces the VHDL concepts which deal with them. These include wait statements and the process sensitivity list.

Chapter 8 continues the examination of sequential modelling by means of a number of examples and introduces the concept of predefined signal attributes as a method of establishing timing relationships.

Chapter 9 builds on the concepts of signal assignment and sequential statement control which were introduced in earlier chapters. It also introduces the concept of concatenation and shows how this can be used in certain modelling situations.

Chapter 10 re.introduces the topic of arrays and discusses intensively the concepts of iterative processes and the use of loops. A method for checking and issuing reports (the ASSERT statement) is also presented in this chapter.

Chapter 11 is concerned with subprograms, explains the applications of procedures and functions within these, and their use in hierarchical design modes.

Chapter 12 deals with types and arithmetic operations and introduces the various types, subtypes, attributes and operators which are defined for the language.

Chapter 13 continues the theme of Chapter 12 and develops examples of the concept of overloading in subprograms and operators. Multiply driven single outputs are dealt with using the resolution function.

Chapter 14 discusses the limits imposed on declarations particularly in packages. Predefined packages are introduced and their relationship to libraries is explained.

Chapter 15 describes the predefined package TEXTIO. The application of the hierarchical modelling technique involving blocks is discussed, as is the idea of guard expressions.

Chapter 16 discusses structural modelling and the idea of non-specific (c.f. algebra) design strategies which can increase the versatility of the written code, e.g. the use of generics to delay decisions on defining parameters such as whether, say, a counter will be 8 bit or 16 bit.

Appendix A is a glossary of terms and is as complete as possible. The 'meanings' given to the words are the authors' and may not be ascribed to the VHDL LRM. Also, the predefined attributes are described in a slightly different way from that of the VHDL LRM.

It is the experience of one of the authors that the material contained in this book can be covered in five two hour sessions by fast-tracking through the Illumination Boxes, a major feature of the book from Chapter 4 onwards. It is assumed, however, that if such an approach were to be used it would need support in the form of workstation exercises, at a ratio of about 3/1 in hours spent at the workstation versus classroom instruction. This procedure, employing one of the many vendors' application tools available, would be an essential feature of the learning process. A more thorough rendering of the information contained in this volume might take some thirty hours in the classroom and, if hands-on workstation or PC instruction were not available, would need to be supported by an equivalent number of tutorial hours.

There are a number of major vendors in this field including:

Cadence Design Systems, Inc.
Comdisco
Compass
Intermetrics
Logic Automation
Mentor Graphics Corporation
Summit Design, Inc.
Synopsis
Vantage Analysis Systems
Viewlogic
Zycad

All of the above vendors provide simulation and synthesis capabilities within their toolsets.

References

1. Ateq Corporation, USA.
2. *Integrated Circuit Engineering,* ASIC Outlook 1994.

Chapter 2

Hardware Design Processes

Overview

This chapter will compare top-down with bottom-up approaches to systems design and trace the trends from the earliest language based attempts at systems modelling through schematic capture techniques to the return towards hardware description languages.

In this chapter you will learn

- Top-down vs bottom-up design approaches
- The comparison between circuit capture and HDL techniques
- Hardware simulation and synthesis processes

2.1 Design Constructs

Any 'system' begins as a design problem to be solved. In many situations, however, similar problems have been previously encountered and the system designer can draw on experience gained from their solution.

Whether the problem to be solved builds on prior experience or constitutes a totally new concept the first requirement will be for a specification. The specification may be precise, detailed and coherently complete or it may be only vaguely articulated. In the first case the designer can proceed to create models confident in the knowledge that the specification is an accurate description of what is required of the system. In the second case model creation, simulation and synthesis will be a recursive procedure carried out in concert with the 'customer' for the system.

Taking the second situation as the more generally realistic, system design may be regarded as a recursive exercise at nearly all levels. It may even be the case that the electrical/electronic elements of the system constitute only a small part of the overall requirement and these will need to be 'fitted-in' to the system specification.

This book is concerned only with the design of electronic digital systems since VHDL was not written with continuous-mode analysis, simulation and synthesis in mind. Several levels of complexity may be identified in the design of digital systems. Of the list given below the last two are not explicitly catered for in VHDL. The design levels are listed in order of abstraction from highest to lowest.

1. System level
2. Algorithmic, or behavioural, level
3. Data flow, or Register Transfer (RTL), level
4. Gate level
5. Switch, or transistor level
6. Electrical, or continuous mode (integral-differential) level

Top-down design (Figure 2.1) will seek to solve the problem by deconstruction from level 1 to level 4, or level 6 (which could include layout), whereas bottom-up design will seek to solve it by building up the system from level 4 to level 1 or, in extremis, from level 6 (full-custom design).

Bottom-up design relies heavily on previous similar experience in that the system must be seen to comprise of standard , or modifiable standard, parts which already exist in some form of library and which can be assembled with confidence. Such an approach may work satisfactorily for small systems but the potential complexity of modern systems makes this method normally non-viable. One of the reasons for this non-viability is the competitive nature of today's markets.

'Time-to-market' has become the critical benchmark for any design project and this period has been falling rapidly in recent years. It is currently between six and eighteen months. Because of this severe constraint it has become necessary for

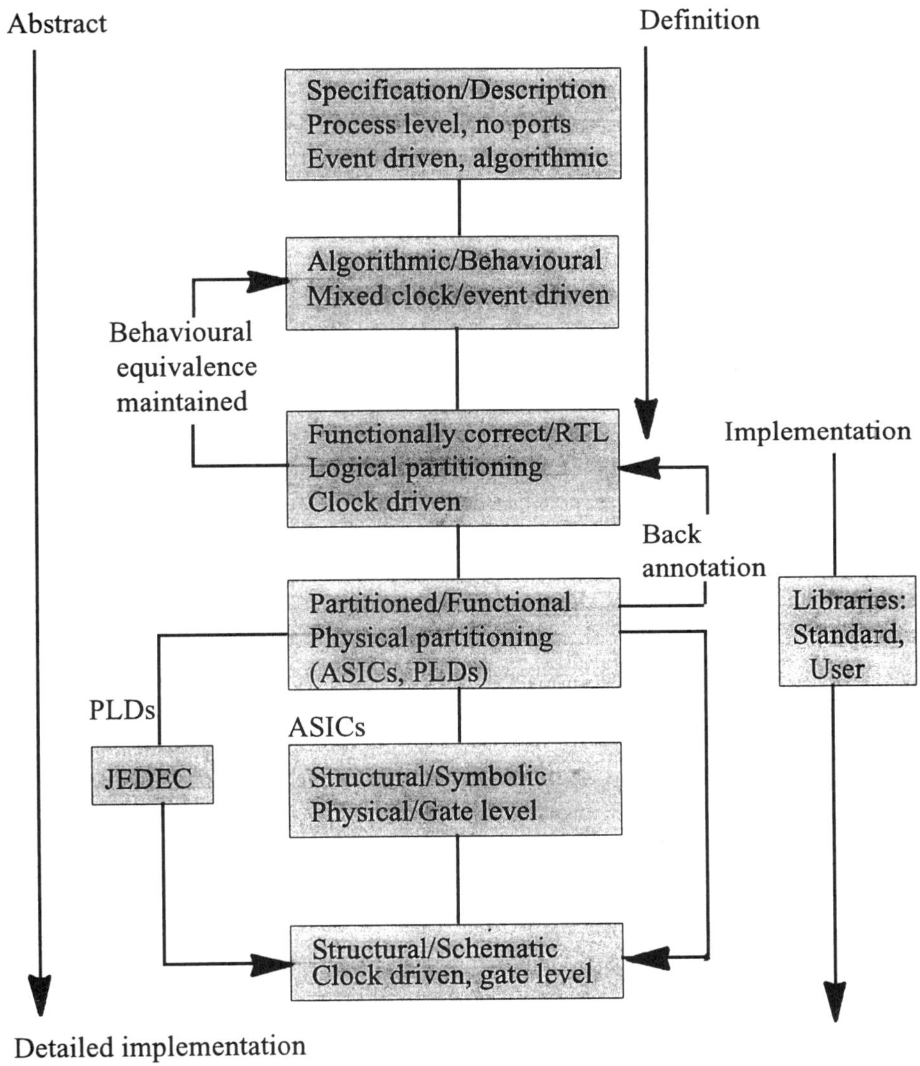

Figure 2.1 Top-down modelling flow[1]

manufacturers to abandon traditional approaches to design and the greatest casualty of the new competitiveness has been bottom-up design strategies. The adoption of top-down strategies has been greatly encouraged by the introduction of powerful computer-aided design tools, not the least of which are the synthesis tools which have emerged in recent years to bring physical reality within the scope of the design and simulation tools which preceded them.

2.2 Design Levels

1. System Level

It is at this level that the system is specified. Furthermore, it is at this level that decisions have to be taken as to how the design should be partitioned. The partitioning referred to here is not the deconstruction of the system into design elements but the decision as to whether the solution will have a software or a hardware bias.

Most 'digital system design problems' respond to microprocessor, or software, based solutions and these may be more, or less, acceptable than hardware (e.g. finite state machine) solutions. What is usually the case is that both solutions may perform equally well and it is often simply a matter of preference as to which is offered. In fact, decisions are, as often as not, a function of company history (and thereby not a little prejudice) as to whether the solution should have a software or a hardware bias.

Only in extreme situations will the problem dictate the method of solution, e.g. where the required response time might preclude a software solution; or the development time, cost and price of the product might dictate a microprocessor programmed approach to the problem. In many cases the 'ideal' solution would be a rational partitioning of the problem between hardware and software and a decision would need to be taken as to where that partition should be placed. Currently this decision has to be a human one since no machine has yet been devised which can make this judgement. Research is, however, under way to produce tools which can generate the most effective solution to system partitioning between hardware and software based on given criteria such as cost, speed, time to market etc.

At the present time tools have been developed to help systems partitioning at the hardware level alone or at the software level alone. Examples of the first kind include Comdisco's Signal Processing Worksystem® (SPW™) and Mentor Graphic's System Design Station® (SDS™). In particular, SDS allows partitioning at the highest abstraction level, that of the specification. It does this by means of schematic signal flow graphs implying state diagrams. These include finite state machine blocks and communication lines to define the relationships between the blocks. Concurrency requirements are handled through synchronisation of the communication lines. These diagrams are decomposable into template VHDL which may then be rewritten to accommodate the actual specification input/output requirements.

At the system level no ports or components are envisaged. This is a 'process' level and each block, or mode, comprises a processor which contains a set of instructions which will allow it to process incoming information and output the results to connected processors in the system. A major characteristic of this level is the production of event driven behavioural code. This level of activity, also known as architectural modelling, is a facility which has only recently been added to the engineer's CAD armoury. It is, therefore, new to digital circuit designers and as such presents a challenge to established approaches to system deconstruction. The system shown in Figure 2.2 below is typical of the type of diagram which can be created using high level design tools. The diagram represents a State Transition Diagram for a stepper controller.

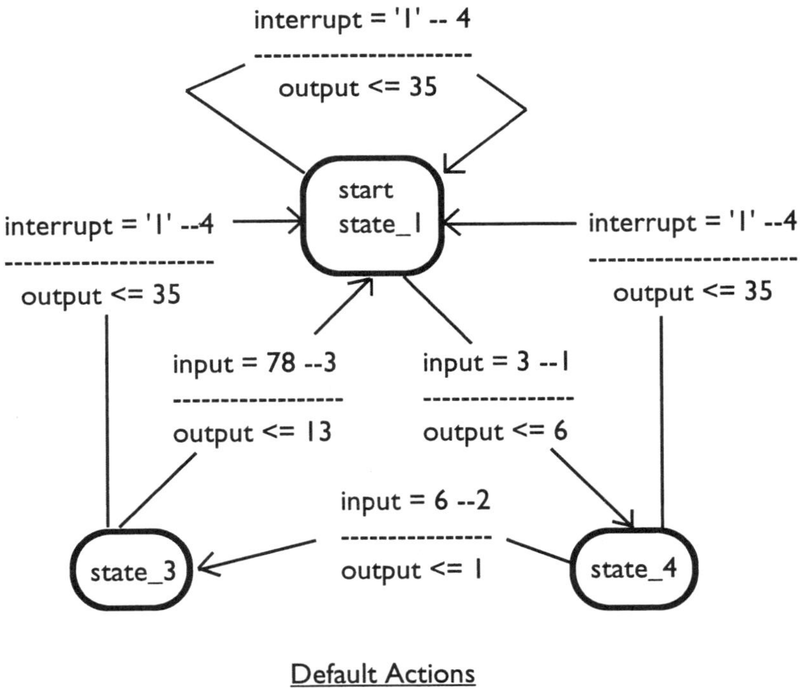

Figure 2.2 State Transition Diagram for a Stepper Controller

2. Algorithmic Level

Below the abstraction of the system level a more defined level may be observed. This level is also known as 'behavioural' and its principal characteristics are that it provides event driven, or mixed clock and event driven, interfaces between the main models of a design; it is behaviourally accurate and its modelling style is preponderantly event driven behavioural code. The example given below in Figure 2.3 is that of the algorithmic level depiction of a simple microcomputer.

At algorithmic level the system may, for example, be described in terms of imperative statements. These are articulated in VHDL in the form of **if-then** statements (see Sections 7.1 and 8.2.1). While the foregoing implies sequential operations it is normally the case that in real systems related events can occur simultaneously. It may also be the case that events may have an equal probability of occurring individually at a particular time. In either case it is necessary that such activity be processed concurrently and, normally, software tools using von Neumann architecture are unable to do this. The ensuing anomaly is taken care of in VHDL by a construct concept known as delta time, a period of time which is defined as being of zero duration (see Section 6.7).

Finally, since real events require time for their completion this is catered for within VHDL by the **after** clause, and alternative outcomes by an **else** (Section 7.1.1) or **elsif** clause (Section 8.3.2). It is a matter of debate at the time of writing as to whether the **after** clause is of much value at the synthesis level. The problem should be resolved when the VITAL (VHDL Initiative Into ASIC Libraries) consortium produces its long-awaited libraries.

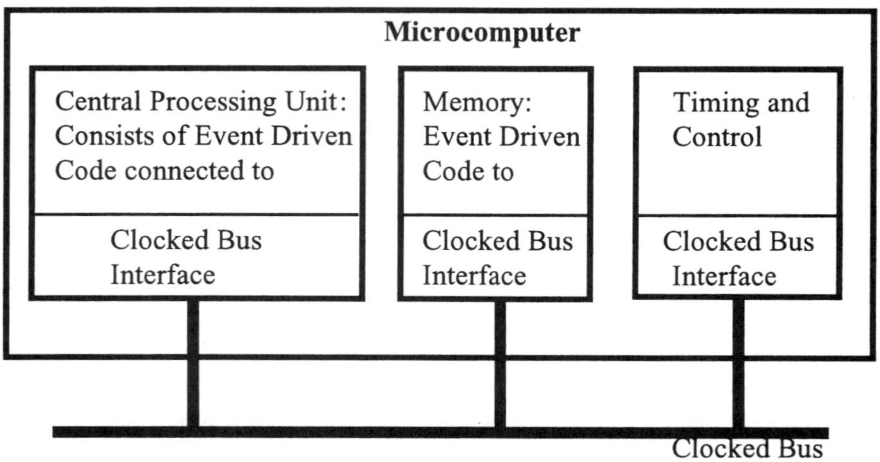

Figure 2.3[1] A Simple Microcomputer

3. Dataflow, or Register Transfer, Level
The main characteristic of this modelling level in VHDL infers fully clock driven Regicter Transfer Level (RTL) code with certain constructs maintained at the behavioural level. RTL modelling embodies the complete functional description and is consistent cycle to cycle. It provides a model which is valid at the next, lower, level i.e. strucural code level. RTL is the best level for simulation in that, because of the logical partitioning implied by the method, simulation will run at its best performance rate. The example given below in Figure 2.4 is for a typical central processing unit as depicted in Figure 2.3 and shows the components defined at RTL.

As the name implies this level deals with the transfer of data and, in particular, the transfer of data between registers. Any processing, or modification, of the data occurs during its transfer. This level of abstraction suggests a very specific mode of operation into which not all digital design tasks can be subsumed. Not only does the level not lend itself to the universal solution of digital questions, it presumes the existence of fairly high level components as the basis of any solution, e.g. shift registers, counters, arithmetic logic units and so on. Furthermore, because of the presence of shift registers and the like the system must, of necessity, be synchronous, depriving the designer of a level of flexibility which might, by the nature of the problem, be needed.

In essence, therefore, although the idea of solving all problems at RTL might be superficially attractive because of its building block approach, the restrictions of the method could, as often as not, outweigh the advantages to the ASIC designer who might well find himself driven either to a higher level of abstraction, if he can afford the tools, or to the next lower level, which is gate level.

VHDL, however, is able to handle this dilemma through the construct of the **package** (see Section 14.5). By this method the required components may be designed at the behavioural level and then be connected at the structural level.

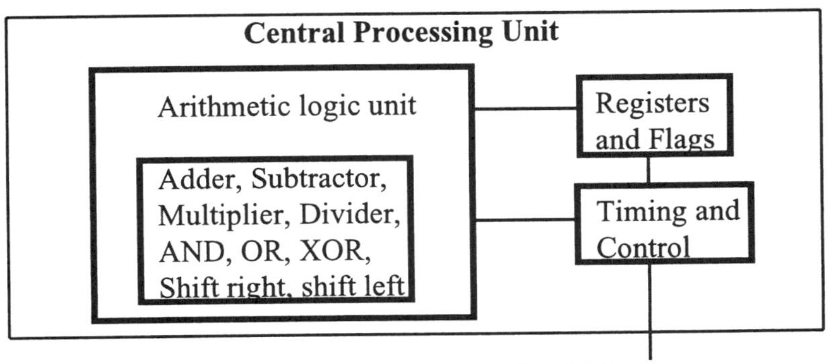

Figure 2.4 A Central Processing Unit

4. Gate Level

Since the introduction of the 74*** series of components some 30 years ago the most popular design approach has been at gate and RTL level. This popularity continued when gate array and cell array design methodology became available to the electronic designer rather than, as hitherto, being the exclusive property of the Silicon Foundry. The development of schematic capture tools in the late 1970s and early 1980s consolidated the position of this design methodology and it has become the cornerstone for nearly all non-Foundry digital circuit design. Switch level design, the next lower level, was, and is, confined to full-custom design methodology. Many design tools did not even provide the capability of observing the circuit at switch level even though this level was more amenable to time domain evaluation through tools such as H-Spice. As a consequence all design is carried out in the value domain (Boolean) and timing problems are addressed at the simulation stage.

In general, CAD tool vendors developed their own simulation methods, enhancing the value domain (limited, within VHDL, to the Boolean levels of '0' and '1') into the electrical domain by adding 'strengths' to the value set, e.g. High Impedance and Indeterminate (and hence taking the function to one remove beyond the Boolean domain).

The real time, or continuous domain, was replaced by a unit delay time model and the internal algorithms of the tools converted the gate level electrical schematic back to a Boolean model for compilation analysis and into an electrical model for simulation. Gates are 'perfect' devices within this methodology, 'rise-time' and 'fall-time' being accounted for by multiples of the unit delay, and most simulation tools make allowance for device imperfection through the concept of min, mean and max delay. Some sophisticated tools make provision for Monte-Carlo analysis of performance.

For VHDL this level implies structural statements only, i.e. simple Boolean statements or **when-else** clauses (see Chapters 5 and 6). The example given in Figure 2.5 shows the gate level implementation of the XOR of Figure 2.4.

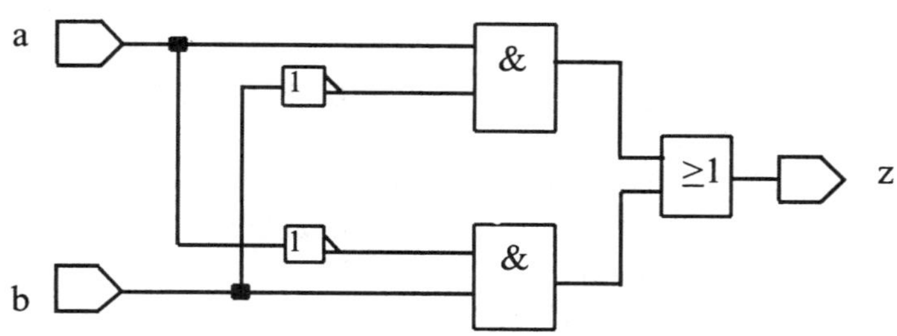

Figure 2.5 Gate-level Implementation of an XOR

5. Switch Level

This level of abstraction tends to be confined to full-custom design, the only concession made to gate level being in the creation of 'in-house' macros. Tools available for this design level include polygon draughting and 'sticks-diagram' modellers. 'Sticks' modelling tools have built into them full-custom, algorithmic, 'polygon pushers' and a manual interactive level. This abstraction level models transistors as ideal on/off switches and the connections between as resistance and discrete capacitance (transmission line modelling is confined to the lower, electrical, level). Thus the modelling level may be continuous in the time domain if required and the value levels have different strengths as well as a logical interpretation.

6. Electrical Level

This is the lowest level of abstraction and is represented by electrical circuits containing resistors, capacitors and inductors. Modelling is done by generating the appropriate integral-differential equations and solving these using transforms such as Laplace and Fourier in the continuous time domain. Non-linearity, or discontinuity, is handled mathematically by transforms such as Heaviside and pulsed waveforms by Z-transforms.

CAD Tools can usually be traced back to the generic, public domain, Spice simulator. Realisation of the product of the design methodology is at the layout (geometric) level and is analogous to full-custom design in the digital environment. Recent developments include Analog Hardware Description Languages (AHDL), e.g. Analogy's SABER™, and the release of an analog extension to VHDL (VHDL-A) is imminent.

2.3 Geometry-based Interchange Formats

One of the earliest attempts at formalising the process of layout for integrated circuits was conceived and developed by Ivan Sutherland and Ron Ayres at Caltech. This is an interchange format, called Caltech Intermediate Form (CIF), which describes the physical features of integrated circuits and lays down a set of rules which can be interpreted independently of any fabrication technology. They can be applied independently of any specific graphics plotting equipment. This independence made the format easy to generate, to transport, and to process and allowed it to become a de facto standard for many university and industrial laboratories during the 1980s. It was, in essence, the first Hardware Description Language which had the virtue of being both machine and human readable, notwithstanding the fact that the language was somewhat terse and confined to the structural level of design.

The development of CIF as a design tool was carried on at Caltech by Carver Mead and Lynn Conway who spread the message of this user method of IC design to other universities such as U.C. Berkeley and M.I.T. Students were now able to design their own integrated circuits and to learn the limitations of the processes involved in their creation using whatever computer and plotting machines were to hand. This was possible by the writing of symbolic language code compatible with the local machines and the generation of translators to CIF. Essentially the students were able, using only primitive design tools, to complete substantial LSI projects and to have these designs fabricated, diced and packaged. Since the chips were multi-project the packages were custom wire bonded for each project and many of the projects proved, through functional testing to perform completely correctly.

The common VLSI system design culture spread and as higher level design aids were developed and shared within this culture, and as standard interface, commercially accessible, implementation facilities were established, more ambitious courses, projects, and research activities were undertaken by students and faculty within the universities[2].

Much of the credit for this expansion must be given to the work of Mead and Conway and their publication[2]. This book led the way for higher education establishments and industrial research units everywhere to pursue development of software and soon there was a flood of formats entering the public domain. These included the Electronic Design Interchange Format (EDIF) and The Electron Beam Exposure System (EBES). Another format, already in existence, called the Gerber Format was adapted and used by many for their own plotter programmes.

As experience was gained and confidence in the methodology increased a number of designers extended the process into formalised, language based, algorithmic descriptions of standard components and their connections. Foremost amongst these was Verilog[3] who developed a Hardware Description Language that was to become the most used text-based method of electronic circuit design during the 1980s. But these methods were used insignificantly by comparison with the almost universal adoption of the circuit capture method of design methodology and it was not until the Department of Defense in the USA commissioned VHDL and compelled its use for DoD contracts that the HDL method was taken seriously. Fortunately the development of HDL methods of circuit design coincided with the almost exponential expansion of the gate counts of new ASIC designs. It was becoming apparent that circuit capture techniques were inadequate for the large gate count requirements of modern systems and designers were driven by necessity to look at the HDL approach to circuit design.

2.4 Schematic Circuit Capture

It soon became clear that circuit capture, using symbols which contained activity (for example, parameters of the Ebers-Moll or Gummel-Poon type at the device level), would speed up the process of IC design since many of the engineers to whom the text based formatting approach was addressed perceived difficulty with a 'software' or text based approach to design of circuits. The main difficulty with a text based approach is the inability to 'see' the circuit being created. Also, in the early days of HDLs, the lack of simulation tools which could read the code rendered this approach less than popular.

The ability to place devices, usually drawn as MIL-STD_806B symbols, and to simulate the result of the placement and interconnections led to an upsurge in the CAD of ICs. The development of computer workstations enhanced this process and software houses were soon offering systems design packages which could be used from the circuit capture stage through simulation and synthesis to layout and sign-off in collaboration with Silicon Vendors. A number of software houses emerged, such as Cadence and Mentor Graphics, and these have come to dominate the CAD scene.

2.5 Hardware Description Languages

As has been previously mentioned, hardware description languages developed in parallel with schematic circuit capture methods. The more powerful of these languages usually owed their origins to an extant software language like Pascal, Fortran or C. By the early 1970s there were some 200 HDLs in circulation and an attempt to harmonise this proliferation was made through the creation of CONLAN, whose 'third generation' successors were the inspiration for VHDL[4]. In the U.K. the preferred HDL was ELLA[5] in that this language had the support of the Ministry of Defence.

The difficulty with a traditional software approach to hardware modelling is that all such languages are von Neumann based i.e. they are sequential, one instruction at a time, processors. It is clear that some aspects of hardware modelling cannot be accommodated by this form of computation. For example, it is not possible to model concurrent circuit behaviour by straightforward sequential computational methods. Hardware modelling languages, therefore, had to find a workaround this problem and the idea of concurrent modelling, i.e. modelling in which the order of statements in code is immaterial, emerged. This requirement is achieved by the construct of zero delay (delta time in VHDL, see Chapter 6).

The other imperative for an HDL is that it must mimic the essential features of the creative process within electronic systems design. Such design processes usually have regard to what already exists in the form of components, sub-systems and the like. Very few designs are created from scratch at the electrical level. Any

HDL which is going to appeal to hardware designers must have features which allow for the assembly of 'parts' and for the creation of these 'parts' in the first place. The idea of creating a combination of parts interconnected by the equivalent of a netlist will be vital to the success of an HDL.

2.6 Hardware Simulation

It is possible, using VHDL, to write a model description which is error free in terms of its syntax but which may not be realisable in terms of intent. The code may, superficially, read as if it would perform the required function but inconsistencies, e.g. timing violations, may be hidden within the description which would not become apparent until the hardware was tested. VHDL code which compiles as error free is not necessarily efficient code or even functionally realisable. Therefore, some method of examining the code before hardware realisation is a necessary condition of VHDL design. A simulator is, in consequence, an inseparable element within the design process.

Simulators are available at all levels of the design cycle. At the top level simulators are used to test the functionality of the system being designed. Simulators for the lower levels provide information on both functionality and timing of the circuits within a design. Figure 2.6 below shows a typical hierarchy for digital system simulators.

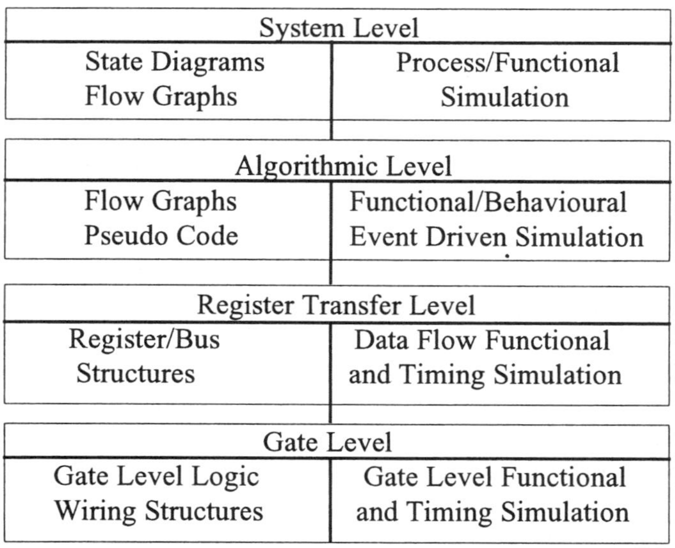

Figure 2.6 Typical Hierarchy for Digital System Simulators

Simulators are offered in two basic forms. The first is a clock driven version in which, after initialisation, all nodes in the circuit elements of a design are swept repeatedly and evaluated at defined timing periods independently of any signal changes on their inputs. The process continues iteratively until the circuit stabilises at which time the next set of input stimuli can be applied. The second type evaluates a circuit element only when at least one of its inputs undergoes a change of state, i.e. there is an event on at least one of its nodes. While this increases the complexity of the simulator it is more efficient in that all nodes unaffected by the input stimulus remain unevaluated. The process algorithm used greatly enhances the speed of simulation and this fact alone renders the event driven simulator the more popular of the two.

2.7 Hardware Synthesis

When a synthesisable model has been 'proved' by simulation the design description which has been created in VHDL can be transformed automatically by a tool described as a hardware synthesiser. There are a number of such tools available commercially and these are more or less efficient in respect of the way in which they deal with the VHDL code supplied to them. The aim of a hardware synthesis tool is to translate the synthesisable description, which has been written in a textual, technology independent HDL, into the most efficient gate-level implementation, in a specified technology and according to predefined criteria such as minimum area, speed etc.

In addition to the above-mentioned externally controllable optimisation features (which are transparent to the user) there are 'hidden' optimisation processes within the tools themselves which perform language-specific optimisations on the design. These include, for example, the removal of redundant code and the simulation of constants to enable their propagation and complete analysis throughout a design.

As mentioned previously in Section 1.3 hardware synthesis tools are becoming more sophisticated and, in generating structure for the design based on the internal representation, they should be capapble of detecting the possibility of resource sharing. Resource sharing is a means of preventing the implementation of multiple copies of circuit elements where these can be demonstrated to be shareable, such as arithmetic operators. Modern tools are able to do this during their internal optimisation sequence.

The synthesis tool must also be able to perform timing analysis for design features such as multiple clocks, cycle sharing, level sensitivity, clock trees and so on. Finally, the modern synthesis tool must be able to deal with the problems of interconnect delays in deep sub-micron (< 0.5 micron) ASIC technologies. They must return the real interconnect delays back to the simulator (back annotation) and not simply estimates of these values. These delays can be measured using the Standard Delay Format (SDF), a timing format which will also be necessary for the implementation of VITAL Library Primitives (Components). SDF will also be used with VITAL to generate timing data for forward propagation to the layout tool.

References

1. With acknowledgements to Cadence Design Systems, Inc.
2. Mead C. and Conway L. *Introduction to VLSI Systems.* Addison-Wesley (1980)
3. Verilog - Cadence Design Systems, Inc.
4. Jean Mermet. *VHDL for Simulation, Synthesis and Formal Proofs of Hardware.* Kluwer Academic Publishers (1992).
5. Royal Signals and Radar Establishment, Malvern, England. *ELLA*.

Chapter 3

IEEE Standard VHDL Language

Overview

This chapter introduces the IEEE Std 1076-1987[1]. It lists the reserved words used by the language and summarises its construction. A design example shows the use of reserved words and identifiers in their context relationships.

In this chapter you will learn

- The VHDL list of reserved words and identifiers
- The structure of VHDL
- Context relationships within VHDL

3.1 VHDL Reserved Words

The VHDL structure is created around a set of lexical elements called reserved words. These words are listed below and are reserved for significance in the language. They are given in lower case bold and this style is used throughout this book in order to facilitate readability.

abs	**else**	**nand**	**select**
access	**elsif**	**new**	**severity**
after	**end**	**next**	**signal**
alias	**entity**	**nor**	**shared***
all	**exit**	**not**	**sla***
and		**null**	**sll***
architecture	**file**		**sra***
array	**for**	**of**	**srl***
assert	**function**	**on**	**subtype**
attribute		**open**	
	generate	**or**	**then**
begin	**generic**	**others**	**to**
block	**group***	**out**	**transport**
body	**guarded**		**type**
buffer		**package**	
bus	**if**	**port**	**unaffected***
	impure*	**postponed***	**units**
case	**in**	**procedure**	**until**
component	**inertial***	**process**	**use**
configuration	**inout**	**pure***	
constant	**is**		**variable**
		range	
disconnect	**label**	**record**	**wait**
downto	**library**	**register**	**when**
	linkage	**reject***	**while**
	literal*	**rem**	**with**
	loop	**report**	
		return	**xnor***
	map	**rol***	**xor**
	mod	**ror***	

Note that reserved words differing only in the use of corresponding upper and lower case letters are considered as the same.

Note also that the reserved word **range** is the only one which has an alternative designation as the name of a predefined attribute.

*Additions to VHDL-1987 in VHDL-1993

3.2 VHDL Structure

1. Design Entities and Configurations.

The design entity is the primary hardware abstraction in VHDL. It represents a portion of a hardware design that has well-defined inputs and outputs (ports) and performs a well-defined function (architecture). A design entity may represent an entire system, a board, a chip, a macro-cell, a logic gate or any level of abstraction in between.

In keeping with normal design practice a design entity may be comprised of a hierarchy, or configuration, of blocks, each of which represents a portion of the entire design. This parallels the concept of symbols in schematic capture tools where such symbols represent progressively higher levels of abstraction in the design. As with such symbols it should be possible to decompose blocks into their structural elements (components) for the purpose of simulation and synthesis at the lowest level of abstraction. This is, however, not a mandatory part of the requirements of the language i.e. it is perfectly possible to generate syntactically correct VHDL which cannot be synthesised; but as an abstraction it may be valid VHDL. For example a design entity may describe a process associated with colour but no synthesis tool may be available to implement the process. File handling, and fault analysis using **assert** and SEVERITY_LEVEL, are other examples.

It should be clear, from the above, that a design entity may also be described in terms of interconnected components. The component itself may be bound to a lower-level design entity in order to define the structure or behaviour of that component. This process is defined in the language as configuration declaration.

Combining the two processes so far described leads to a hierarchy of design entities which will represent a complete design.

All of the above is reminiscent of a process whereby components are connected together to create a design and where, normally, a library exists which contains associated data which may be drawn upon to test the functionality of the components and the subsequent system.

2. Subprograms and Packages

Subprograms define algorithms for computing values or exhibiting behaviour. They can be used to perform computations and are a means of resolving multiple drivers of a common signal. They may also be used to define portions of a process. Packages provide a means of defining the above such that different design units can share the same declarations (see Chapter 14).

There are two forms of subprogram: procedures and functions. A procedure call is a statement; a function call is an expression and returns a value, e.g. in the design example in Section 3.4 there is a requirement for 8 identical counters and 8 identical comparators. A package could therefore be used to declare the procedure for the counter and the function of the comparator.

The subprogram format is similar to that of the design entity in that it can comprise two parts: a subprogram declaration which defines its calling conventions, and a subprogram body defining its execution. Packages may also be defined in two parts: a package declaration to define the visible contents and a body to provide hidden details.

3. Types

A type is characterised by a set of values and a set of operations. There are four classes of types:

a. Scalar types: integer, floating point, physical and enumeration.
b. Composite types: array, record.
c. Access types: provide access to objects of a given type.
d. File types; provide access to objects which contain a sequence of values of a given type.

4. Declarations

There are many different forms of declaration, each of which associates an identifier with a declared entity. A declaration takes effect through the process of elaboration which is, essentially, the execution of the model either through simulation or synthesis or both.

5. Specifications

Specifications relate to entities which already exist. Thus a specification associates additional information with a previously declared entity. It follows, therefore, that a given specification must either follow, or be contained within, the declaration of the entity to which it relates.

There are three kinds of specification:

a. Attribute.
b. Configuration.
c. Disconnection.

6. Names

Names are used to denote declared entities whether declared explicitly or implicitly. They can also denote objects denoted by access values and subelements, or slices, of composite objects and values. They may also denote attributes of any of the foregoing.

7. Expressions

An expression is a formula which defines the computation of a value.

8. Sequential Statements

Sequential statements are used to define algorithms for the execution of a subprogram or process. They execute in the order in which they appear.

9. Concurrent Statements

Concurrent statements are used to define interconnected blocks and processes which jointly describe the overall behaviour or structure of a design.

Concurrent statements execute asynchronously with respect to each other. The primary concurrent statements are the block statement, which groups together other concurrent statements, and the process statement which represents a single independent sequential process. A block statement defines an internal block representing a portion of a design. Blocks may be hierarchically nested to support design decomposition. A process statement defines an independent sequential process representing the behaviour of some portion of the design.

10. Scope and Visibility

Scope may be described as the range of the identifier name appearing in the declaration. It is the region of the model where it may be possible to link an identifier name to the meaning given to the name in a particular declaration. This area extends from the beginning of the declaration to the **end** in the region within which the declaration appears.

Because regions in VHDL occur within other regions the scope of identifiers tends to follow a hierarchical structure.

Visibility refers to the linking of identifier names with particular declarations. A declaration is said to be "visible" when an identifier name receives its meaning from that declaration. The declaration is "hidden" when the identifier name does not associate with that particular declaration.

11. Design Units

Design units are constructs which may be independently analysed and inserted into a design library. One or more design units in sequence comprise a design file. Design units in a design file are analysed in the textual order of their appearance in the design file. Analysis of a design unit defines the corresponding library unit in a design library. A library unit is either a primary unit or a secondary unit. A secondary unit is a separately analysed body of a primary unit resulting from a previous analysis.

12. Elaboration and Execution.

Elaboration is the process by which a declaration achieves its effect. In order to execute a model the design hierarchy defining the model must first be elaborated. This is followed by the initialisation of the nets in the model. Finally, simulation of the model can proceed. Thus the elaboration of a design hierarchy will create a collection of processes interconnected by nets and this collection of processes and nets may then be executed to simulate the behaviour of the design.

13. Lexical Elements

The text of a description consists of one or more design files. The text of a design file is a sequence of lexical elements each composed of characters.

14. Predefined Language Environment

This consists of the predefined attributes and the packages STANDARD and TEXTIO that all VHDL implementations must provide.

3.3 A Design Example

The following example is designed to demonstrate the correlation between the VHDL description of a system and the equivalent schematic capture version.

It should be understood that at this point in the book the reader is not expected totally to comprehend the code which follows, simply to recognise the context relationship between the implementing code and the circuit elements which a hardware designer might invoke and connect together in order to fulfil the specification.

Suppose that the following specification for a nuclear radiation monitoring system has been supplied:

A scintillation detector is required to count bursts of radiation which will be represented by a three bit signal stream. It is necessary to discriminate between the strengths (eight strength levels) and the number of pulses of the various burst strengths which may be encountered and so a threshold limiter must be included in the design. The threshold limiter must be able to set the burst count at between 1 and 16 pulses, i.e. 4 bits, for each strength level of the input. The detector must be able to indicate which strength level has reached the preset threshold first and to reset the associated counter, this threshold level having been reached.

Let us first create a block diagram of the proposed solution from the point of view of a hardware designer.

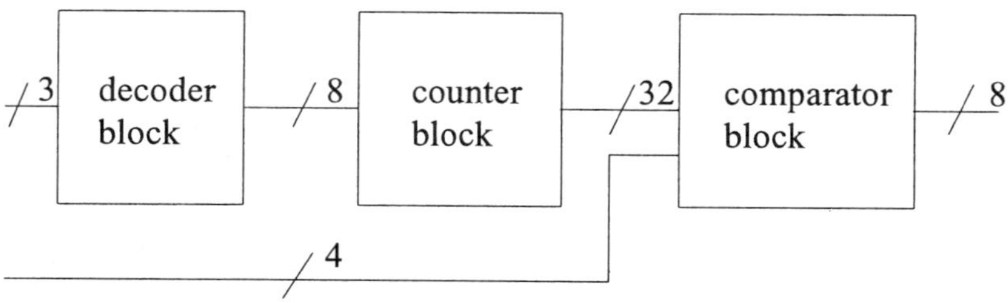

It should be clear from the above diagram that we need a 3 to 8 decoder, 8 counters, and 8 comparators. It would be a simple matter, therefore to decompose the above blocks into gate level circuitry to implement the blocks e.g. the decoder will require, at the simplest level of implementation, eight 3-input and_gates each having the three input signals or their inverse applied to these gates in accordance with the truth table for three inputs.

It should also be recognised, however, that a designer may have other design criteria to which he must respond e.g. smallest area, or best performance, which would inform his decisions as to whether he should try to minimise the gate count (a hardware designer will be aware that the design of this decoder could be minimised for area by using two input or_gates where appropriate).

The counter block will comprise eight 4-bit counters which are started by the condition of the decoder outputs. On the assumption that the counter enable is active low then the signal from the decoder which will begin the count for any strength signal will be '0'. The outputs from the counters will pass to comparators which will compare these outputs with a preset threshold level. The first comparator to demonstrate that the threshold has been reached will produce an output signal which may be sent to an indicator and, also, could be use to reset the associated counter.

The hardware designer might represent the decoder as shown in Figure 3.1 overleaf. The decoder has signal inputs *x0* , *x1* , *x2* and signal outputs *y0* through *y7*. In VHDL these input and output signals may be described by defining them as ports in a port declaration as follows (see Chapter 4):

```
port ( x2 , x1 , x0 : in BIT ; -- declares xn to be mode in
    y7, y6 , y5 ,  y4 , y3 , y2 , y1 , y0 : out BIT) ; -- and of
                                                -- type BIT.
```

NB. the character (--) is used to append comments to the code for clarification. These comments do not form part of the VHDL code and are discounted by any compiler. Thus the words after the characters above may be read as the sentence "declares xn to be mode **in** and of type BIT."

The signals could also be described in terms of their obvious groupings using the VHDL Package STANDARD Predefined Array Type BIT_VECTOR (see Appendix A):

```
port ( x : in BIT_VECTOR ( 2 downto 0) ;
    y : out BIT_VECTOR (7 downto 0)) ;
```

Note that the type BIT_VECTOR does *not* describe a logical word of length *n*; rather it identifies a group of similar, but individual, bits each of whose attributes is related to, and corresponds with, its position in the group.

Having described the input/output relationships of the decoder it is now necessary to define how it functions or behaves.

In VHDL this process is controlled by the reserved word **architecture** and defines how an entity behaves within the boundaries of the port declarations. Suppose the part (entity) has been called simply 'decoder'. Then, one possible architecture might describe the structure of the decoder and could be written thus:

```
architecture structure of decoder is
-- declarative statements
begin
-- activity statements
end structure;
```

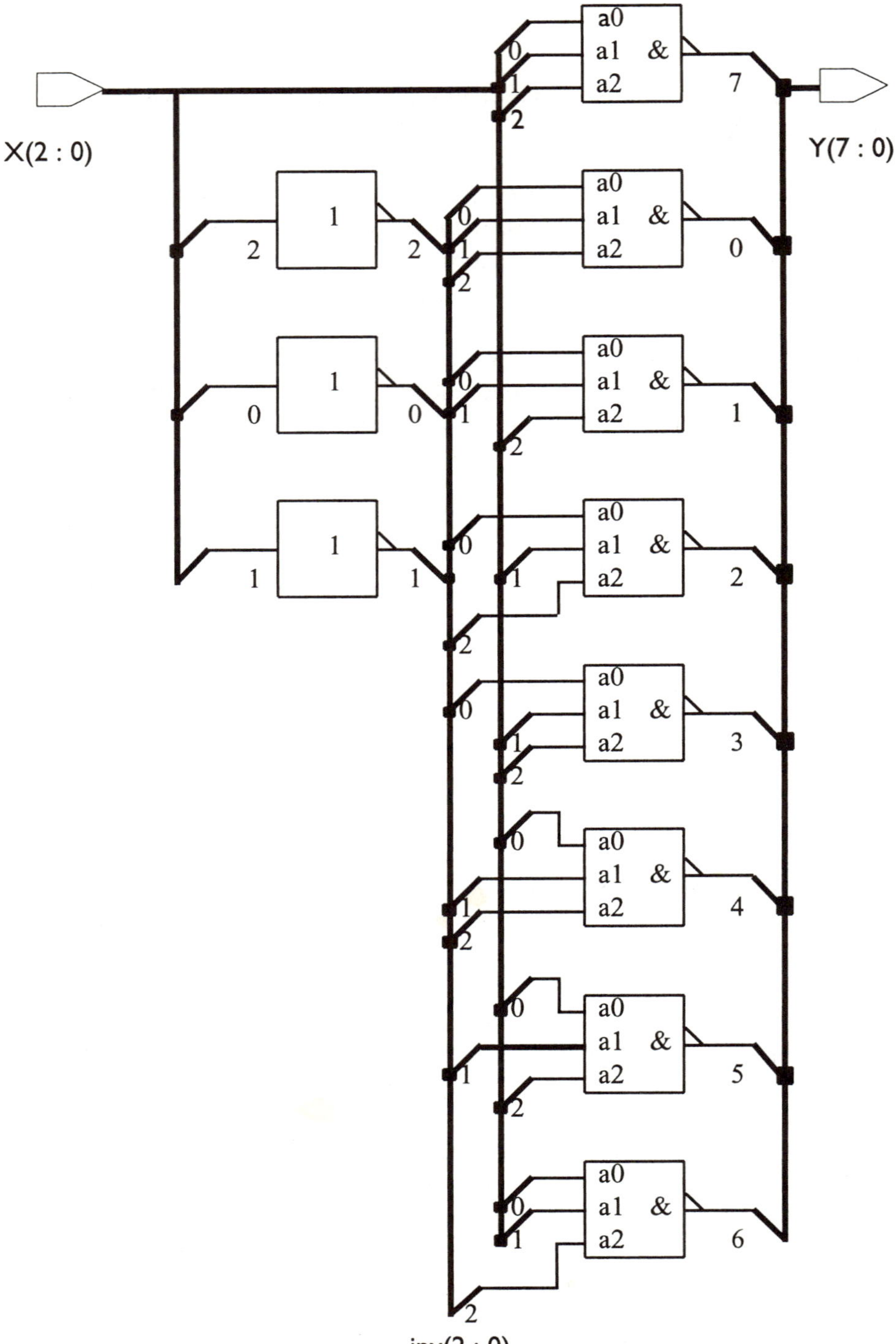

Figure 3.1 Schematic diagram of 3-8 Decoder

It may be the case, however, that an alternative description for the decc involving its behaviour, could be written (see Section 6.1). This would imp., need for a different name in order to discriminate this description from the previous one, e.g.:

```
architecture behaviour of decoder is
-- declarative statements
begin
-- activity statements
end behaviour;
```

Both of these descriptions are valid for the entity and it may be thought prudent to include both in the description of the entity. However, only one of these can be activated at any time and so some method must exist whereby the correct version can be invoked for a particular application, such as simulation.

The following are examples of valid descriptions for the activity of the decoder:

```
------------------------------------------------------------------------------
--    Example 1
------------------------------------------------------------------------------
        entity decoder_1 is
        port ( x2 , x1 , x0 : in BIT ;
            y7, y6 , y5 , y4 , y3 , y2 , y1 , y0 : out BIT) ;
        end decoder_1 ;

        architecture structure_1 of decoder_1 is
        begin
        y0 <= '0' when ( x2 = '0' and x1 = '0' and x0 = '0') else '1' ;
        y1 <= '0' when ( x2 = '0' and x1 = '0' and x0 = '1') else '1' ;
        y2 <= '0' when ( x2 = '0' and x1 = '1' and x0 = '0') else '1' ;
        y3 <= '0' when ( x2 = '0' and x1 = '1' and x0 = '1') else '1' ;
        y4 <= '0' when ( x2 = '1' and x1 = '0' and x0 = '0') else '1' ;
        y5 <= '0' when ( x2 = '1' and x1 = '0' and x0 = '1') else '1' ;
        y6 <= '0' when ( x2 = '1' and x1 = '1' and x0 = '0') else '1' ;
        y7 <= '0' when ( x2 = '1' and x1 = '1' and x0 = '1') else '1' ;
        end structure_1 ;
------------------------------------------------------------------------------
--    Example 2
------------------------------------------------------------------------------
        entity decoder_2 is
        port ( x : in BIT_VECTOR ( 2 downto 0) ;
               y : out BIT_VECTOR (7 downto 0)) ;
        end decoder_2 ;
```

```
            architecture structure_2 of decoder_2 is
            begin
y   <=      "11111110" when x = "000" else
            "11111101" when x = "001" else
            "11111011" when x = "010" else
            "11110111" when x = "011" else
            "11101111" when x = "100" else
            "11011111" when x = "101" else
            "10111111" when x = "110" else
            "01111111" when x = "111" else
            "11111111" ;
end structure_2 ;
```

Note that the operator <= appears only once and that double quotes are used for arrays (BIT_VECTOR) as compared with single quotes for a single BIT. Note also that closing quotes only are used, in compliance with VHDL requirements. This style is used throughout the book for consistency in the presentation.

Another way to describe the decoder is as an assembly of directly connected nand_gate components (a purely structural description). The gates themselves might have multiple architectures as follows:

```
entity nand3_gate is
port (a0, a1, a2 : in BIT ;
      b : out BIT) ;
end nand3_gate ;

architecture logical_1 of nand3_gate is
begin
b <= '0' when a0 ='1' and a1 = '1' and a2 = '1'  else '1' ;
end logical_1 ;

architecture logical_2 of nand3_gate is
begin
   process (a0, a1, a2)
   begin
         if a0 = '1' and a1 = '1' and a2 = '1' then b <= '0' ;
         else b <= '1';
         end if;
   end process;
end logical_2 ;
entity inverter is
   port (f : in BIT ;
```

```
g : out BIT) ;
end inverter ;
architecture invert of inverter is
begin
    g <= not f ; -- declares Boolean behaviour of inverter
end invert ;
entity decoder is
port ( x : in BIT_VECTOR ( 2 downto 0) ;
        y : out BIT_VECTOR (7 downto 0)) ;
end decoder ;

architecture structural of decoder is
component nand_3
    port (a0, a1, a2 : in BIT ;
            b : out BIT)) ;
end component ;
component inverse
    port (f : in BIT ;
            g : out BIT) ;
end component ;
for nand0, nand1, nand2, nand3, nand4, nand5, nand6, nand7 :
nand_3
-- calls for 8 copies of nand_3 component
use entity WORK.nand3_gate (logical_2) ;
-- uses architecture logical_2 of entity nand3_gate
for inv1, inv2, inv3 : inverse use entity WORK.inverter ;
signal inv : BIT_VECTOR (2 downto 0) ;
--carries signals from outputs of inverters to inputs of nand_gates.
-------------------------------------------------------------------------------------
-- the following port mappings represent the interconnections between the
-- components. The parameters map from the formals in the components and
-- to the actuals in the entities. The mappings are by relative position.
begin
    inv1: inverse port map ( x(0), inv(0) ) ;
    inv2: inverse port map ( x(1), inv(1) ) ;
    inv3: inverse port map ( x(2), inv(2) ) ;
    nand0 : nand_3 port map ( inv(0), inv(1), inv(2), y(0) ) ;
    nand1 : nand_3 port map ( inv(0), inv(1), x(0), y(1) ) ;
    nand2 : nand_3 port map ( inv(0), x(1), inv(2), y(2) ) ;
    nand3 : nand_3 port map ( inv(0), x(1), x(2), y(3) ) ;
```

```
        nand4 : nand_3 port map ( x(0), inv(1), inv(2), y(4) ) ;
        nand5 : nand_3 port map ( x(0), inv(1), x(2), y(5) ) ;
        nand6 : nand_3 port map ( x(0), x(1), inv(2), y(6) ) ;
        nand7 : nand_3 port map ( x(0), x(1), x(2), y(7) ) ;
    end structural ;
```

Note that this description does not use the *values* of the signal BITs or BIT_VECTOR to generate the design (except for the description of the nand_gate itself). Note also the use of the word WORK. This word defines a library into which all entities are placed by default, i.e. it exists implicitly and the entities within it can be called upon in the way demonstrated. The syntax from **for** to **end** structural is called the configuration specification for the decoder (see Section 16.5).

The description given above is essentially a straightforward statement of the way in which the components comprising the decoder would be connected together. It makes no use of any power available in a compiler to interpret an algorithmic or behavioural description of a 3 to 8 decoder, as was evinced in the earlier versions of the entity. Clearly, therefore, one would not use such a cumbersome method to describe an entity as simple as a decoder. Nevertheless, because it is a simple entity, the method serves to illustrate the concepts of hierarchical VHDL design using components and signal mapping.

The remainder of the design will be considered in future chapters of this book and completed designs are given (in several different interpretations) in the Exercises Solutions Manual which accompanies this book

Having introduced VHDL by means of reserved words, identifiers, and context relationships the rest of the book is devoted to taking a detailed look at the many aspects of the language.

Finally, it should be stressed that, while the final illustration of the application of VHDL to the design of a decoder used in this chapter relied heavily on the ability to *see* the schematic implementation of the design to help the development the appropriate code, the reader is encouraged to abandon this approach in favour of the far more powerful method of thinking only in terms of a language, or text based, description of the problem to be solved. Such descriptions may, of course, include truth tables, function tables or state transition tables as well as, for example, state diagrams, flow diagrams, Mealy and Moore models and so on.

Reference

1. *IEEE Standard VHDL Language Reference Manuals* (IEEE Std 1076-1987 and ANSI/IEEE Std 1076-1993).

Chapter 4

Modelling with VHDL

Overview

Model creation is a process. At some point there must be an idea of what is to be modelled. Sometimes this might be a model of an existing device; in other cases a specification of the relationship between the inputs and outputs of a device may be all that exists. Even this specification may not be all that specific, but it must exist in order to create a starting point for the modelling.

In this chapter you will learn

- How to create an entity from a specification
- To relate a simple analysis of a design into a model structure
- How to write an entity declaration and a port statement
- The meaning of the architecture body

4.1 Model Development

Specification

Analysis

Design

Entry

Test

Compile

Debug **Debug**

Simulate

4.1 Model Development

Model creation is a process. The steps listed in the Illumination Box 4.1 are one approach to the model development process.

1. Specification: What is the problem?

   At some point there must be an idea of what is to be modelled. In some cases this may be a model of an existing device; in other cases someone may ask for a device which, given a certain set of inputs, will produce a certain set of outputs. The specification may be very general or very specific, but a specification must exist.

   In most cases the better the specification the easier it will be to create a model which will perform like the device performs or is required to perform.

2. Analysis: What is the solution?

   Another way to express this is to ask the question "What is the relationship between the input and the output"?. The modeller must think about this problem. Techniques such as truth tables and flow-charts can be used to represent the relationships between inputs and outputs.

3. Design: How is the solution expressed in VHDL?

   What structures in VHDL can be used to implement the solution arrived at in step 2? How do certain solutions translate into VHDL? In short, this is the coding phase of the model's development.

4. Entry: What is necessary to prepare the model for testing?

   A model must be entered into a testing system before it can be tested. Entry is more than simply keying the model into a file. Stimulus tests must be devised and, in some cases, entered into stimulus files.

5. Test: Does the model do what it is supposed to do?

   As indicated by the graphics in the Illumination Box 4-1 testing is an iterative process. The code must be compiled. If syntax errors occur it will be necessary to correct them and recompile. Once the model compiles without syntax errors it must be stimulated to test for logic errors and design errors. If either of these errors are discovered the code will have to be changed and the testing process begun again.

This development process may be repeated many times in the creation of a single model. It would be used first at a very high level of abstraction to show "proof of concept". It would be used again as refinements break the model into smaller logical units and add structure to the design.

4.2 Specification

Create an entity which receives two digital signals and which outputs a single signal. If both input signals are low the output signal is to be high. For any other combination of high or low inputs the input is to be low.

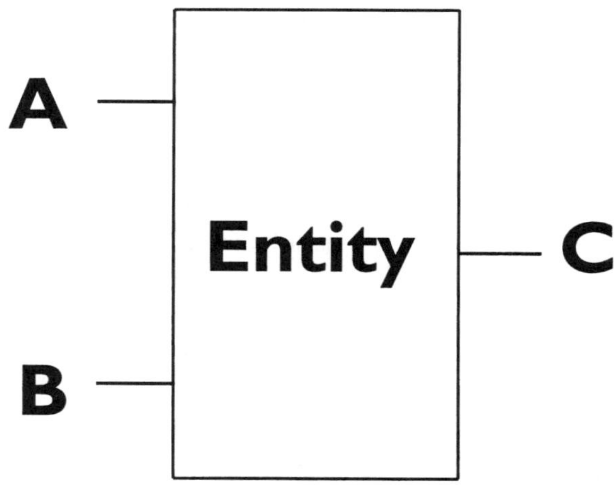

4.2 Specification

Let us suppose that a specification has been generated which requires that a device be created which receives two digital signals and which outputs a single signal. Furthermore, if both input signals are low the output signal is to be high. For any other combination of high or low inputs the output is to be low.

This simple specification, it will be recognised, represents a NOR gate and already the reader will have, in his mind's eye, a symbol for the device; but we are concerned with the *process* of modelling the device textually rather than symbolically and not, therefore, on the creation of a complex algorithm for the model.

However, there is no reason, at this stage, why the symbolic constructs of the hardware oriented modeller should not be used to draw parallels between symbols and words when modelling in VHDL. This will enhance and encourage the transition towards thinking in Hardware Description Language terms if symbols are shown with the VHDL structures used to represent them.

4.3 Analysis:

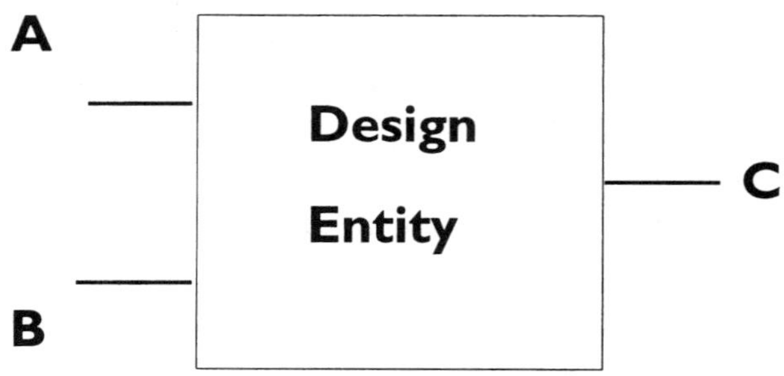

Truth Table

Inputs		Output
A	B	C
0	0	1
0	1	0
1	0	0
1	1	0

- **Basic modelling unit is called a Design Entity**

- **A Design Entity may represent an entire system, a board, a chip or a gate**

4.3 Analysis

Once given a specification, the modeller will begin to provide definition to the problem through analysis of the design. Truth tables, flow charts and other means of expressing logic constructs may be used.

Analysis is usually more complex than simply creating a truth table. A truth table shows what output to expect with what inputs; it does not show *how* to achieve the output given the input. Flowcharts and other methods are better for showing the "how". This first model, therefore, will be constructed as a data-flow or register transfer level (RTL) model. By using a truth table our focus will not be distracted by the introduction of abstract logic constructs.

4.3.1 Modelling levels

The three levels of modelling in VHDL are:

1. Algorithmic or Behavioural
   This is similar in form to Behavioural Language Modelling (BLM) but not as complicated to construct. Usually, models of this type consist of long segments of sequential code. Generally, if an **if** statement appears in the code there is a good chance that the model can be classified as algorithmic.
2. Data-flow or RTL
   Models of this type show the assignment of values to outputs based upon the value of the inputs. Although implicit **if**s occur explicit **if**s are not stated.
3. Structural
   Structural modelling is much like creating a net list. Parts which have been described at one of the two other levels can be instantiated and connected verbally to form a more complex model.

4.4 Design Entity

It should be noticed that the label on the symbol in Illumination Box 4.3/4 has been changed from "ENTITY" to "DESIGN ENTITY". In VHDL a model is referred to as a "design entity". This "design entity" may represent a gate, a chip, a board or even an entire system.

At the beginning of a design process the design entity will probably model the entire system; or a substantial part of the system. This is done to show proof of concept. As the design project progresses the design may be broken down into a series of smaller functional units which can be modelled to show proof of concept at this lower level. Later, these units may be broken down into still smaller units and so on down to gate level (VHDL does not easily model at the silicon level).

4.5 Model Structure

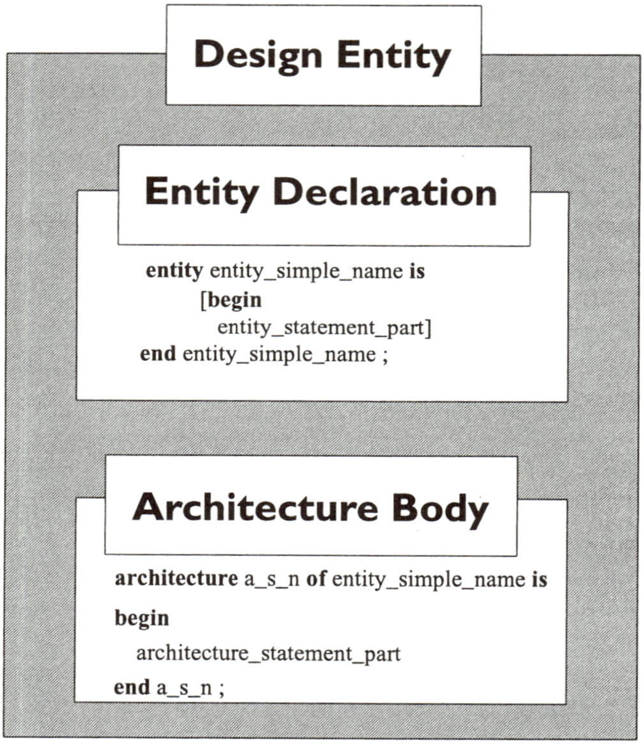

- **VHDL Models consist of an Entity Declaration and an Architecture Body**

However, at each point, the fundamental unit of functionality which is being modelled is referred to as the design unit.

At this point the "general" modelling capabilities of VHDL must be stressed. Throughout much of this book the examples are limited to simple electronic devices. This is done to maintain a level of familiarity which should make the reader feel more comfortable with the processes. It is enough to introduce the fundamentals of the language without having to appreciate complex design structures. However, this approach misses a very important aspect of VHDL; The ability to model very abstract functionality. This aspect is dealt with in a limited way in this volume (Chapter 9).

4.5 Model Structure

Whilst there are few limitations upon what a "design entity" may be there are certain rules as to what VHDL constructs may go into making up a design entity.

Illumination Box 4.5 illustrates the model structures which comprise a VHDL model. There are two of these, the Entity Declaration and the Architecture Body. Both of these constructs can become very complex but, for the moment, our task is to keep them simple. In a later chapter (Chapter 7) we will add complexity to these issues.

Design entities must have two parts in order to be classified as complete;

1. An Entity Declaration:
   The Entity Declaration provides the entity with a name and defines the interface between the entity and its environment. There can be one, and only one, entity declaration for a design entity.

2. An Architecture Body:
   The Architecture Body establishes the relationship between inputs and outputs within the design entity. There may be more than one architecture body associated with an entity declaration; however, one, and only one, architecture body may be active at a time (unless the design entity has been instantiated several times in a design and then only one architecture body for each instantiation may be active).

4.5.1 entity Declaration

- **Names entity and defines interface between entity and its environment**

- **Format**

```
entity entity_name is
      [port (port_list) ;]
end [entity_name] ;
```

- **Example**

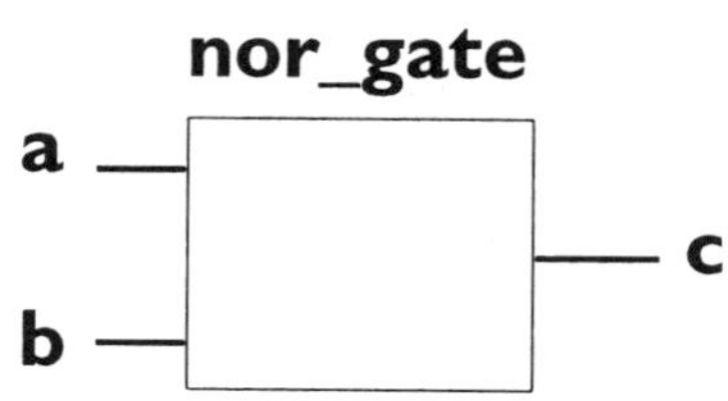


```
entity nor_gate is
      port (a, b : in BIT;
                  c : out BIT) ;
end nor_gate ;
```

4.5.1 entity Declaration

The Illumination Box 4.5.1 illustrates the format of an entity declaration and gives an example.

The Format shows an abbreviated form of the entity declaration.

The following conventions will be observed when presenting the format of VHDL structures:

1. Lower case
   Words which appear in lower case will, in general, be simple_name identifiers
2. Lower case bold
   Words which appear in **lower case bold** are VHDL reserved word identifiers. Their form cannot be altered. They must appear as written. They must not be used as explicitly declared identifiers.
3. Brackets
   Brackets [ ] are used to indicate optional parts of the VHDL structure. Structures within brackets may be omitted.
4. Parentheses
   Parentheses ( ) are part of the VHDL structure and must appear as indicated.
5. Words appearing in CAPITALS are predefined types or attribute identifiers.

For example, when describing the format of the entity declaration the entity name after **end** is optional; however, it is recommended that it always be included for documentation purposes and clarity when reading a model.

The entity name is indicated by "entity_name" in the Illumination Box 4.5.1. This name is the name given by the modeller to the model and can be any name the designer wishes (except that it cannot be a VHDL reserved word). The rules for creating these names will be discussed in Chapter 5 (Section 5.1.2). For the moment it may be assumed that any name composed of letters (alpha characters) and underscores may be used given the proviso mentioned above.

It may be noticed that the **port** clause is optional. Whilst this is true, a model created without a **port** clause has no way of accepting data from outside itself and no way of transmitting data beyond itself. This would severely restrict the usefulness of such a model. However, one example of a model without a **port** clause is to be seen in the creation of test benches for models, for example:

```
entity TestBench is
end TestBench ;
```

The issues of test benches and their relationships to the creation of test vectors, simulation, the collection and analysis of test results, and the validation of designs are not dealt with explicitly within this text but are to be found within the solutions

4.5.2 port Statement

- **Identifies ports used by the entity to communicate with its environment**

- **Format**

  port (name_list: mode type;
  •
  •
  •
  name_list: mode type) ;

- **Example**

  nor_gate

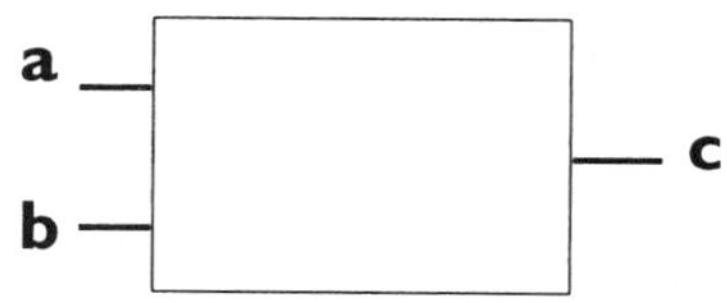

port(a, b **: in** BIT **;**
c **: out** BIT**) ;**

to the Exercises in the Solutions Manual.

The example at the bottom of the Illumination Box 4.5.1 shows how the entity declaration for the specification of the nor_gate in Illumination Box 4.4 could be constructed. Notice that a name has been given to the entity and that the ports to the entity are described. The Illumination Box 4.5.2 discusses the port clause in more detail.

4.5.2 port Statement

A "port" may be thought of as an entry into or an exit from an entity.
The **port** statement in the entity declaration identifies ports used by the entity to communicate with its environment.

Objects identified implicitly in **port** *clauses are signals.*

While the **port** statement is optional an entity without this statement will have no way of communicating with its environment. No signals can be input to the entity and no signals can exit from the entity.

The **port** statement comprises the word **port** followed by a list of the port names, their mode (i.e. whether they conduct signals into, or out of, the entity), and the type of signal (identifying what values the signal may assume) and ended by a semicolon (;).

Statements in VHDL end with a semicolon. Thus a single statement can be "spread out" over several physical lines without any indication of continuation. Look at the example of the **port** statement at the bottom of Illumination Box 4.5.2. Notice the section a, b : **in** BIT;. The a, b is an example of a *name_list.*

All port names which appear in a single list must have the same mode and type.

In this example the ports *a* and *b* both conduct data into the entity and, in both cases, the type of data is 'bit' (1 or 0). The semicolon at the end of a, b : **in** BIT; does not indicate the end of the **port** statement; instead it indicates the end of that particular name list and to the fact that another name list follows.

The last name list in this **port** statement is c : **out** BIT. The port *c* has a different mode from the other two ports and, therefore, must have its own name list. It is not followed by a semicolon since no other name lists follow it.

4.5.3 port Mode

- Identifies direction of data flow through port

- All ports must have an identified mode

- Allowable modes

in	-	flow is into entity
out	-	flow is out of entity
inout	-	flow may be either in or out
buffer	-	in or out limited to one 'in' source
linkage	-	documentation only does not pass signal information cannot be read from or written to

4.5.3 port Mode

The "mode" of a port identifies the direction of data flow through the port.

All ports must have a mode identified in the port clause.

There are five possible port modes in VHDL

1. **in**
   Data can enter the entity through ports of mode IN but may not exit.

2. **out**
   Data can exit the entity through ports of mode OUT but may not enter.

3. **inout**
   Data may either enter, or exit, through ports of this mode.

4. **buffer**
   Data may either enter, or exit, through ports of this mode. However, buffer ports differ from inout ports in that they must never have more than one source. This is due to the fact that signals of type buffer cannot be resolved signals. Therefore, signals of type buffer may only be connected to non-resolved local signals or to other buffer signals.

5. **linkage**
   The primary purpose of signals of type linkage is documentation. Signals of this type cannot be used to pass information. They cannot be read from or written to, therefore they do not specify a direction of data flow.

4.5.4 Architecture Body

- **Establishes the relationship between inputs and outputs of a design**
- **Format**

```
architecture body_name of entity_name is
        declarative_statements

    begin
        activity_statements

    end [body_name];
```

- **Example**

```
architecture data_flow of nor_gate is

begin
    activity_statements
end data_flow;
```

4.5.4 Architecture Body

The architecture body of the model establishes the relationship between the inputs to the entity and the outputs from the entity. Another way to express this might be "given this input, here is how you arrive at this output". The architecture body contains the steps, or process, needed to arrive at the correct output, given a certain input. It may be expressed in terms of structure, data flow, or behaviour.

To be a complete "design entity", a model must have an entity declaration and at least one architecture body.

NB. An entity declaration, without an architecture body, may be placed in a file and compiled. This is also possible for an architecture body if an entity declaration has already been created and compiled. Note that the entity must always be compiled before the architecture and that, if an entity is recompiled, all architectures associated with it must be recompiled.

An architecture statement has three major features:

architecture body_name **of** entity_name **is**

names the architecture and associates it with a given entity declaration. The <body_name> is chosen by the modeller.

The <entity_name> is the name of the entity with which the architecture is associated and must match exactly the name given in the entity declaration.

NB. The name of the architecture becomes important when multiple architectures are associated with a single entity declaration. In those cases the names must be unique.

2. **begin**
   begin separates the declarative area of the architecture from the activity area. The activity area contains the statements which process the input to arrive at the output.

3. **end** [body_name];
   Although inclusion of the body_name is optional it is recommended that it always be used for documentation and readability purposes.

4.5.5 Commenting Code

- **A double hyphen (--) indicates that all material from that point to the end of the line is to be treated as comment**

- **Example**

```
Architecture example of comments is
            -- area before begin
            -- is called declarative area
      begin
            -- area after begin contains
            -- statements which have activity
      end example;
```

4.5.5 Commenting Code

Referring to Illumination Box 4.5.5 it will be noticed that the declarative area and the activity area in the architecture statement are indicated thus:

-- declarative statements

-- activity statements

A double hyphen (--) in VHDL indicates that all the material from that point on the line to the end of the physical line is to be treated as a comment. If a line is split between statement and comment the comment must follow the statement.

Comments are parsed out of the code during compilation. Therefore, while a very long model with many comments might take slightly longer to compile, the simulation time of models which are identical except for the number of comments should be the same.

It is recommended that comments be used throughout a model. A well commented model is easier to understand and to change at future times.

4.5.6 nor_gate Architecture Body

Truth Table

Inputs		Output
a	b	c
0	0	1
0	1	0
1	0	0
1	1	0

```
architecture data_flow of nor_gate is
begin

   c<= '1' when a = '0' and b = '0' else
        '0' when a = '0' and b = '1' else
        '0' when a = '1' and b = '0' else
        '0' when a = '1' and b = '1' else
        '0' ;
end data_flow;
```

4.5.6 nor_gate Architecture Body

Illumination Box 4.5.6 introduces the "activity" which produces the result shown in the truth table.

The body_name "data_flow" has been chosen because of the modelling style used in this example. The name nor_gate was used in the entity declaration and must, therefore, be used for the architecture body. There are no declarative statements in this model and so nothing appears between the **architecture** and **begin**.

Statements in the "activity" area cause changes in signals. In this model the statement assigns a value to the output port based on the values that appear on the input ports. For example, when the value of signal *a* is '0' and the value of signal *b* is '0' the value '1' is assigned to port *c* If *a* and *b* are not both '0' then the next condition is checked and this process continues until a match is found or until a value to be assigned is encountered which has no conditions. Note that the last line of a signal assignment statement cannot contain a condition. This is necessary in order to provide a "default" value assignment when no match exists in the list of conditions.

In this example the signal assignment statement was written exactly to match the truth table; however, this is not necessarily the most efficient way to write the signal assignment statement and it might be preferable to note that since "1' is assigned only when *a* is '0' and *b* is '0', and all other assignments are '0', the signal assignment could have been written:

```
c <= '1' when a = '0' and b = '0' else
     '0';
```

Also, as will be seen later, certain approaches to signal assignment may not produce the "best" synthesis of the model and that "work arounds" may be necessary in order to achieve the most desirable implementation.

Note: The symbol <= used for the signal assignment operation (See Section 5.1.3) is the only symbol which may be used for this purpose. The signal assignment operator has a hardware connotation and therefore implies a non-zero time relationship between the assigned signal waveforms and the signal assigned to.

4.5.7 nor_gate Model

Design Entity

Entity Declaration

```
entity nor_gate is
        port (a, b : in BIT;
              c : out BIT);
end nor_gate;
```

Architecture Body

```
architecture data_flow of nor_gate is
begin
c <= '1' when a ='0' and b = '0' else
     '0' when a = '0' and b = '1' else
     '0' when a = '1' and b = '0' else
     '0' when a = '1' and b = '1' else
     '0';
end data_flow;
```

4.5.7 nor_gate Model

It has now been shown how to construct an entity declaration and an architecture body. Illumination Box 4.5.7 shows a complete model of the device described in the earlier specification.

Note that:

1. The model consists of an entity declaration and an architecture body.

2. The architecture body is linked to the entity declaration by the repetition of the entity name in the architecture body.

3. The entity declaration declares the interface between the entity and its environment.
   In this case the entity has two signals entering it. The value of the signals entering the entity are limited to '1' an '0'; similarly for the value leaving the entity.

4. The relationship between the two input signals and the output signal is described in the architecture body.
   The signal assignment assigns a value to the output signal based on the values of the two input signals.
   Note that the symbol use to assign to signals is <=.
   Note also, in the example given opposite, that the symbol <= is used once only during the assignment to *c*, i.e. the assignment is not a group of separate statements. The right-hand assignment comprises a single statement (and could, therefore, be written in a single line but which is presented as seen for clarity) .

4.6 Exercises

4.6.1 Describe the purpose of
i) The Entity Declaration
ii) The Architecture Body

4.6.2 The following model code was produced to represent a 2-input NAND gate. It contains a number of code errors. What are they?

```
entity nand gate is
    port (a, b ; in BIT
              c ; out BIT) ;
architecture data_flow of nand gate is
begin
c    <=     '0' when a = '1' and b = '1' else
     <=     '1' when a = '0' and b = '1' else
     <=     '1' when a = '1' and b = '0' else
     <=     '1' when a = '0' and b = '0'
end nand gate ;
```

4.6.3 Produce a design entity for a 3-input and_gate and include commenting code where appropriate.

4.6.4 Produce a model for a '2-to-4' decoder.

4.6.5 Implement the simple switching function $f = C(A + B)$ in VHDL code.

4.6.6 Write a dataflow description for a one bit 'half-adder'.

4.6.7 The truth table given in Figure 4.1 is for one of the four 2 to 1 multiplexers of the 74LS157. Write a dataflow description for one multiplexer.

ENABLE	SELECT INPUT	INPUTS		OUTPUT
E_bar	S	I0	I1	Z
H	X	X	X	L
L	H	X	L	L
L	H	X	H	H
L	L	L	X	L
L	L	H	X	H

H = HIGH Voltage Level
L = LOW Voltage Level
X = Don't Care

Figure 4.1 Truth Table

Chapter 5

Objects, Data Types and Operators

Overview

VHDL is a hardware modelling language designed to be vendor-independent. Nevertheless it must model reality. Real elements with real parameters must be reflected within the described models. This chapter will introduce the structural dimensions necessary to parameterise the data flow descriptions of Chapter 4.

In this chapter you will learn

- The concept of objects
- How to assign values to objects
- How to perform Data Type operations on objects
- The meaning of Logical Operators and Overloading
- The concept of Delay

5.1 Objects

Objects in VHDL are containers for values.

There are four classes of objects in VHDL

1. **Constants**
2. **Variables**
3. **Signals**
4. **Files***

NB. The variable class of objects also has an additional subclass: shared variables

*Added to VHDL-1993

5.1 Objects

Objects in VHDL are containers for values.

There are three classes of objects in VHDL:

1. Constants
   Constants are objects whose values, once having been assigned, cannot be changed. They are "constant" value sources. Constants can be declared in any declarative area in a VHDL model and may be used throughout the model.
2. Variables
   Variables are objects which can be assigned a value that may be changed during the simulation of a model. A variable is a container in which a single value can be stored. There are restrictions as to where variables may be declared and used in VHDL models i.e. only within "sequential" areas of the model. Also includes the additional subclass: shared variables VHDL 1993 LRM, Section 4, 4.3
3. Signals
   Signals are objects which can be assigned values and an associated time factor for each value. Although a signal may assume only one value at a time, it has an associated driver which may contain projected, or future, values (values having an assigned time at which the value may, or will, take effect). Signals can be used throughout the VHDL model but may be declared only within the "concurrent" areas of the model.
4. Files (Added to VHDL-1993)
   Files are objects whose values are the sequence of values contained in the host system file. File types are used to define objects representing files in the host system environment.

The assignments of values to variables and signals have some differences which may produce markedly different outcomes. These differences, and their outcomes, will be discussed after the declaration and use of variables is introduced. For the moment, signals are the only objects which will be used in models.

Hardware designers may find it useful to equate "signals" to physical wires. A wire is a physical device used to transport a value from one point to another. In the same way a "signal" may be thought of as a "device" for the transport of a value from one point in the model to another. Since, in real digital processing systems and in their models, values are simply symbols of reality this is a true and useful analogy.

5.1.1 Declaring Objects

- **Declaration format:**

  Object_Class identifier : **type** [:= init_value];

- **Examples**
  - **Constants**

    constant delay : TIME := 10 ns;
    constant multiplier : REAL := 5.25;
    constant prop_del : DELAY_LENGTH -- a
    -- deferred constant

  - **Variables**

    variable sum : REAL ;
    variable voltage : INTEGER := 0;

  - **Signals**

    signal clock : BIT;
    signal inner_works : Std_logic := 'X';

  - **Files**

    type file_type **is file of** type_mark ;

- **Objects in a port clause are classified as signals by default**

5.1.1 Declaring Objects

It has not been necessary to make an explicit declaration of objects in models so far because:

Objects in the **port** *statement of an entity declaration are classified as signals by default*

cf.. the wire analogy given above.

It should be remembered, however, that objects used in a model *must* be declared, whether they are declared explicitly in an object declaration statement or implicitly in a **port** statement.

The general format for declaring objects is:

OBJECT_CLASS identifier : **type** [:= init_value];

The OBJECT_CLASS is either **constant, variable** or **signal**. The identifier is the name supplied by the user for this object. **type** identifies the values which may be assigned to the object.

The phrase ":= init_value" is optional. If it is included it represents the initial value to be assigned to the object.

Declaration of objects can only take place in declarative areas within the model. These areas exist through out the VHDL structure but some classes of objects may only be declared in certain areas. The only declarative area so far introduced is the one which appears just before the **begin** in the architecture body viz.:

```
architecture example of declarative is
            -- declarative area
begin
            -- activity area
end example:
```

Only constants and signals may be declared in this area. Illumination Box 5.1.1 show examples of objects used in declarative areas.

5.1.2 Naming Objects

- **Valid characters**
  - alpha characters (a - z)
  - numeric characters (0 - 9)
  - underscore (_)

- **Must start with an alpha character***

- **May consist of any number of alpha, numeric, or underline characters**

- **Underscore must be preceded and followed by alpha or numeric characters**

- **Names are case insensitive**

* **VHDL-1993 allows extended identifiers, which may contain any graphic character, if enclosed within backslashes (See Section 3.2)**

5.1.2 Naming Objects

In the VHDL language all user supplied names must follow certain naming conventions. These naming rules include the following:

1. Names can be constructed using the following characters;
   - alpha characters - the letters of the English alphabet.
   - numeric characters - the characters 0, 1, 2, 3, 4, 5, 6, 7, 8, 9
   - underscore - the underscore character (_).

   NOTE: Proprietary VHDL CAD tools may include character extensions to the VHDL set.
2. Names must start with an alpha character.
3. Names may consist of any number of alpha, numeric or underline characters.

   NOTE: There may, however, be physical limits imposed by a system editor; for example 'a name cannot be longer than one line'. In some editors this is 256 characters. Common sense also dictates that, although VHDL imposes no upper limit on the length of an identifier (name), any name over, say, 25 characters would be difficult to remember and to reproduce.
4. Underscores must be preceded, and followed, by alpha or numeric characters.

   NOTE: Two underscores cannot be placed together, i.e. the name al_pha is legal, but the name al__pha is illegal.
5. Names are case insensitive.

   Since VHDL is a case insensitive language a word may be written in upper case letters, lower case letters or a mixture of both. For example the name 'play' is recognised as one and the same object whether it is written PLAY, play, PlAy, pLAy or any other combination; the case of the letters makes no difference in VHDL.

5.1.3 Signal Assignment

*	Multiplying operator - multiplication
**	Miscellaneous operator - exponentiation.
/	Multiplying operator - division.
+	Adding operator, addition, plus
-	Adding operator, subtraction, minus.
&	Adding operator. Used for concatenation, e.g. linking two or more items into a single unit. Items must be of the same type.
=	Relational operator. Equal
<	Relational operator. Less than.
>	Relational operator. Greater than
/=	Relational operator. Not equal.
<=	Relational operator. Less than or equal.
>=	Relational operator. Greater than or equal.

<= The only assignment operator which can be used to assign waveforms to signals. The appropriate use of this operator is decided by the context of its application.

5.1.3 Signal Assignment

The assignment of a waveform value to a signal is referred to as signal assignment. There are certain rules to be observed when assigning values to signals. These are as follows:

1. The value assigned to a signal must be one of the values matching the "type" of the signal.

NOTE: A "type" associates a name with a set of values. For example, **bit**, which is the only type which has, so far, been introduced. The values associated with type **bit** are '0' and '1'.

2. The single quotes around the values used in the assignment are mandatory. These values are referred to as "character literals" and *are* case sensitive. The character literals 'X' and 'x' are not the same value.

   Character literals must appear in single quote marks, must consist of a single character and case sensitivity must be acknowledged.

3. The less than/equal to symbol (<=) is the only operator which can be used to assign waveform values to signals.
4. Signal assignment statements, like other VHDL statements, are terminated with a semicolon. The statement itself may extend over several physical lines as, for example, in the conditional assignment statement in Illumination Box 4.5.4 which extends over five physical lines but which represents a single VHDL statement.

5.2 Initial Value Assignment

- **Value assumed by object at time of declaration**
- **Appears after type in declaration**
- **Declaration format:**
  object_class identifier **type** [:=init_value];
- **Example**
  variable voltage : INTEGER := 0 ;
- **If not assigned, system assumes a default value equal to left-most or minimum value for type**

5.1.4 Conditional Signal Assignment

While the static signal assignment has limited usefulness the conditional signal assignment has wider application. It was used in the model example in Chapter 4. Conditional signal assignment assigns waveform values to signals based on the validity of a condition.

Conditional signal assignment structure:

1. The waveform value associated with a **when** condition is assigned to the signal if the condition following the **when** produces a "TRUE".
2. There must be at least one "**when** condition **else**" in the statement.
3. The last waveform value in the conditional signal assignment statement cannot have a **when** condition associated with it. This value is the "default" value and is assigned only when none of the conditions following **whens** in the statement evaluate as "TRUE".
4. In a statement which contains multiple "**when** condition **else**" constructs:
   · The conditions are evaluated in the order in which they are written.
   · The value associated with the first condition which produces a "TRUE" is assigned to the signal.
   · No other conditions are checked after the first to evaluate as "TRUE". It is immaterial whether subsequent conditions are "TRUE" or "FALSE".

NOTE: multiple assignment operators must not be used with conditional signal assignment statements. These will cause errors. For example:

```
signal_a  <=   '1' when b = '0' else
          <=   '0';
```

The second (<=) is illegal.

Conditional signal assignment statements must contain only one assignment operator.

5.2 Initial Value Assignment

An object may be assigned an initial value when it is declared. This value will be the value assigned and used in the evaluation when a simulator is first invoked on the component. This process may be necessary where the default value set may have a different order from that required in the simulation. For example; the default order of type BIT is '0','1', the left most value being invoked first.

The initial value must be consistent with the type of the object viz:

```
signal example_signal : BIT := '1';
```

is a valid statement because the type BIT consists of the value set '0' and '1'.

5.3 Data types

- Identify set of values an object may assume and operations which may be performed
- Classifications
  - Scalar
    Numeric, enumeration, and physical objects
  - Composite
    Arrays
    Records
  - Access
    Value sets which point to dynamic variables
  - File
    File types are external files of values and provide access to objects that contain a sequence of values of a given type

If an initial value is not explicitly assigned to an object when it is declared, VHDL will assign a default value equal to the left most, or minimum, value of the type used in declaring the object. In the example above had it been declared as follows:

signal example_signal : BIT;

then the initial value assigned to the **signal** <example_signal> would be '0' as the left most value in BIT's value set.

It is worth reminding the reader at this point that the **port** statement implicitly declares signals (cf.. 5.1.1). For example:

port (a, b : **in** BIT;
c : **out** BIT);

declares three signals a, b and c which, since no user supplied initial value has been assigned to them, must take the initial value '0'. On the other hand, the following statement:

port (a, b : **in** BIT := '1';
c : **out** BIT);

has assigned the initial value of '1' to signals "a" and "b" but has assigned the initial value '0' to "c" by default.

5.3 Data Types

It should now be clear that the type of an object identifies the values which can be assigned to it. In addition, the type also identifies the operations which may be performed on the object.

There are four classifications of types:

1. Scalar
   Scalar types may be further broken down into
   · numeric - types whose value sets are either integer or floating point numbers
   · enumeration - types whose value sets are either characters literals or identifiers
   · physical objects - types whose value sets can be broken down into physical units
2. Composite
   Composite types may be further broken down into:
   · arrays - types consisting of multiple elements of the same type
   · records - types consisting of multiple elements each element of which can be of a different type

5.3.1 Scalar Types

- **Numeric**
  Type INTEGER **is range** -2147483647 to 2147483647;
  Type REAL **is range** -1.0E38 to 1.0E38;

- **Enumeration**
  type BIT **is** ('0','1');
  type BOOLEAN **is** (FALSE, TRUE);

- **Physical**
  type TIME **is range** -2147483647 to + 2147483647

```
    UNITS
    fs;
    ps  = 1000 fs;
    ns  = 1000 ps;
    us  = 1000 ns;
    ms  = 1000 us;
    sec = 1000 ms;
    min =   60 sec;
    end UNITS;
```

3. Access

   Access types have value sets which are pointers to objects of some other type.
4. File

   File types are external files of values and provide access to objects that contain a sequence of values of a given type

In this book discussion will be limited to scalar and composite types. Access and file types are beyond the scope of this introduction to VHDL.

5.3.1 Scalar Types

Although, at this stage of the discussion, it is not necessary to be able to create type declarations it is important to understand how they establish a value set. Illumination Box 5.3.1 illustrates some examples of scalar type definitions.

5.3.2 Scalar Examples

1. Numeric

   · **type** INTEGER **is range** -2147483647 **to** 2147483647;

   This type declaration associates the name integer with integer num bers from -2147483647 to 2147483647. Because neither has a decimal point we know that the values in the set are integer numbers. However, until the name integer was linked to this set of symbols by this type of declaration it had no relationship to integer numbers. The IEEE Language Reference Manual for VHDL states that the range of integer can be implementation dependent. This means that propri etary implementations of VHDL may support a larger or smaller range than the one given above (which is the range of Integer in the IEEE package STANDARD), i.e. the range is hardware dependent.

   · **type** REAL **is range** -1.0E38 to 1.0E38

   This type declaration associates the name real with floating point numbers from -1.0E38 to 1.0E38*. Because these numbers contain a decimal point we know that the values in the set are floating point numbers.

   *The range of real is also allowed to be implementation (hardware) dependent (IEEE LRM).
2. Enumeration

   · **type** new_bit **is** ('X', '0', '1');

   This type declaration associates the name new_bit with the values 'X', '0' and '1'. Values written in this manner are called character literals. While VHDL is a case insensitive language character literals *are* case sensitive. Hence, 'X' is not the same enumeration type as 'x'.

   · **type** colour **is** (red, yellow, green);

5.3.3 Predefined Types

- **Certain scalar data types are predefined and do not require a type definition statement**
- **Examples**
  - **BOOLEAN**
  - **BIT**
  - **INTEGER**
  - **REAL**
  - **CHARACTER**
  - **TIME**
  - **SEVERITY_LEVEL**

This type declaration associates the name colour with the identifiers red, yellow and green. Identifiers are user chosen names. Names used as identifiers are case insensitive.

3. Physical
   Physical types specify a value range and a base unit. They may also specify secondary units which must be multiples of the base unit. The value range of physical types must be an integer range. The maximum allowable range will normally be the same as that of the integer implementation range of the proprietary tool employed.
   NOTE: Although TIME is a predefined physical type it does not have range limits defined by implementation. TIME is a special case and will always have the maximum range specified in IEEE LRM. This range is unique to TIME and it cannot, therefore, be used in user defined physical types. It should be appreciated, however, that the time range used in a simulator can differ from the range TIME residing in the system's compiler.

5.3.3 Predefined Types

Certain scalar types are defined in a package called STANDARD. This package is defined as part of the IEEE standard and is automatically included in every model (a USE clause for this package is implicitly assumed to exist at the beginning of the model. See Section 5.3.5)

Type declarations which appear in package STANDARD have the effect of being predefined types and do not require a type declaration statement.

The following scalar types and the indicated value sets appear in the package STANDARD.

1. BOOLEAN
   value set: FALSE, TRUE.
2. BIT
   value set: '0', '1'.
   (NB. BIT_VECTORs are enclosed in double quotes, " ").
3. INTEGER
   value set: -2147483647 to 2147483647
4. REAL
   value set: -1.0E38 to 1.0E38
5. CHARACTER
   value set: Extensive; all the upper and lower case letters, the numerical symbols, certain special characters and control characters. The IEEE Std 1076-1993 contains the complete list (See Section 3.2).

5.3.4 Non-STANDARD Predefined Types

- **IEEE.Std_logic_1164 recognises a signal type composed of 'X', '0', '1' and 'Z'**
- **Rather than writing type definitions for this type in every model it would be convenient to have it predefined**
- **Sharing of type definitions is done through constructs known as packages**
- **Common IEEE definitions and functions are located in IEEE supplied packages such as Std_logic_1164**
- **'use' statements make packages "visible" to a model**
- **Example**

```
use IEEE.Std_logic_1164.all ;
```

6. TIME
   value set: -2147483647 to +2147483647. The base unit is the femtosecond (fs).
7. SEVERITY_LEVEL
   value set: NOTE, WARNING, ERROR, FAILURE

It should be noted that even though type definitions for each of these types exist in the package STANDARD, it is possible to create a type definition within a model which assigns a different value set to any of these names. However, creating type definitions which "re-define" these type names should be avoided if only to prevent confusion. Which value set was being used would not always be intuitively clear and might cause inexplicable logic errors. It is recommended that these names be treated as if they were reserved words.

In order to emphasise this recommendation predefined types will be shown in capital letters throughout this book.

5.3.4 Non-STANDARD Predefined Types

If we recall that one of the primary objectives for the commissioning of VHDL was to facilitate the design of electronic circuits it might come as a surprise to observe that, in the predefined type BIT, the writers of the language made no provision for the realities of indeterminance, high impedance in nets or differently strengthed binary signals.

Many digital simulators take into account the possibilities of signals having different strengths e.g. whether a '1' has been derived from the power source (V_{dd} say) or by restoration. Because such types are commonly used in simulators it would be useful to predefine these rather than have to write a type definition for such types into every model. Writing a type definition into many models presents several sources of possible error, for example, the initial value of an object, if not explicitly stated, is the left most value in the object's type value set. Thus, if the value set is written ('X', '0', '1', 'Z') the initial value would be 'X'; however, if the required initial value were '0' this might produce a difference in model execution from that expected by the modeller (See sections 13.2.4/6).

To avoid this, and other problems, type definitions can be shared through constructs known as packages. In fact, this is how the type definitions specified in the IEEE LRM standard are shared. No modification is allowed to the IEEE package STANDARD and so where types other than those contained in STANDARD are required for common use these must be created in user defined packages. For example the IEEE has created a package IEEE Std 1164-1993, with 9 states, 'U', 'X', '0', '1', 'Z', 'W', 'L', 'H' AND '-', which every VHDL vendor must include. Unlike the package STANDARD such packages are not implicitly included in models and for these to be included it is necessary to invoke a LIBRARY statement

5.3.5 Accessing User Predefined Packages

- **Std_logic can be used in models if the package Std_logic_1164 is included in the model**

- **Example**

```
library IEEE ;
use IEEE.Std_logic_1164.all ;

entity xor_gate is
        port (a, b : in Std_logic ;
                   c : out Std_logic) ;
end xor_gate ;

architecture data_flow of xor_gate is
begin
c<=   '0' when a = '0' and b = '0' else
      '1' when a = '0' and b = '1' else
      '1' when a = '1' and b = '0' else
      '0' when a = '1' and b = '1' else
      'X' ;
end data_flow ;
```

and a USE statement which must be placed at the top of the model file. The USE statement makes the contents of a package visible within the model.

For example; the statement

use IEEE.Std_logic_1164.**all;**

makes <u>all</u> of the contents of the package "Std_logic" located in the library "IEEE" visible within the model in which it appears.

Packages will be discussed in greater detail in Chapter 14.

5.3.5 Accessing User Predefined Packages

The architecture of the nor_gate of Illumination Box 4.5.7 suggests that the gate is 'ideal'. It does not allow for the possibility that the inputs a and b, and therefore the output c, might be indeterminate. These conditions are possible in real circuits and therefore the possibilities need to be accounted for and identified. The identifier 'X' (see 5.3.3[2]) might be used to represent the indeterminate state and is, in fact the enumeration type used by IEEE within type Std_logic, whose value set is ('X', '0', '1', 'Z'). The package which contains Std_logic, as has been previously mentioned, is Std_logic_1164. Therefore, in order to avail the model of these values the following statements must appear at the top of a model file:

library IEEE;
use IEEE.Std_logic_1164.**all**;

NOTE: The **library** statement provides the logical name used to point to the location of the desired package. Notice also that the name of the library must appear both in the **library** statement and in the **use** statement. The **.all** indicates that all of the definitions and declarations in this package are to be linked to this model.

Finally, it would not have been possible to start the model declaration by simply using

port (a, b: **in** Std_logic;

because the VHDL compiler would have no pointer as to the location of Std_logic, or what values were associated with it, without reference to the appropriate library.

The model of the xor_gate in Illumination Box 5.3.5 may now be compared with the nor_gate model of Illumination Box 4.5.7. The model of the former is closer to reality than the latter. The default value at the end of the conditional signal assignment has been changed from '0' to 'X'. Since 'X' is a value in the Std_logic value set, this is an allowable change. It is saying that if the values on the input are not '1' or '0' the model cannot determine the value of the output.

It is interesting to consider the effect of this approach on the test values for simulation. Since the input ports are now also Std_logic it is possible to force either, or both, to 'X' or 'Z'. The result, in this model, would be that the output port would change to 'X'.

5.4 Operators:

- **The type of an object also identifies the operations which may be performed on the object**

- **Operators defined for certain predefined data types:**

- **Logical operators**
  and | or | nand | nor | xor | xnor

- **Relational operators**
  = | /= | < | <= | > \ >=

- **Adding operators**
  + | - | &

- **Sign**
  + | -

- **Multiplying operators**
  *** | / | mod | rem**

- **Miscellaneous operators**
  **** | abs | not**

Introduction to VHDL

5.4 Operators

The type of an object identifies the values which may be assigned to the object *and* the operations which may be performed on the object. Operations are identified by operators. These may be classified into the following groups:

1. Logical Operators ::= **and** | **or** | **nand** | **nor** | **xor** | **xnor** | {**not**}
   It should be noted that **not** is classified as a logical operator which has the precedence of a miscellaneous operator and, therefore, does *not* appear in the list of logical operator lexical elements in the LRM of the IEEE. However, since logical operators are presented in this Volume in terms of the equivalent gate operation, and since **not** acts somewhat like an inverter, it may be placed in this category without giving rise to ambiguity. These operators are defined for the predefined types BIT and BOOLEAN and for one dimensional arrays of type BIT and BOOLEAN. NB. **xnor** was added to VHDL - 1993.
2. Relational operators ::= = | /= | < | <= | > | >=
   These operators test for ordering. The objects used on either side of these operators must be of the same type. These operators are defined for *any* scalar type. The result of comparisons using relational operators is always of type BOOLEAN.
3. Adding operators ::= + | - | &
   These operators perform "classic" arithmetic operations. They are defined primarily for numeric and physical types. Which numeric types can be used with which operator is one of the subjects of Chapter 10. The sign & indicates concatenation (the joining) of two objects into one one-dimensional array. It will be discussed in detail in Chapter 9.
4. Sign ::= + | -
   The + and - are unary sign operators and are defined for numeric types. They represent the identity and negation.
5. Multiplying operators ::= * | / | **mod** | **rem**
6. Miscellaneous operators ::= ** | **abs** | **not**

It will have been noticed that the relational operator <= is identical to the already introduced signal assignment operator <= and that the addition operators are identical to the sign operators + and -. Whilst it might seem that this duplication would cause confusion it should be recognised that VHDL checks the context, or usage, of the symbol in order to decide its purpose during simulation of the model.

5.4.1 Logical Operators

- **Predefined types for types BIT and BOOLEAN**
- **Follow conventional definitions**
  - **and**
  - **or**
  - **nand**
  - **nor**
  - **xor**
  - **xnor**
  - **not**

5.4.1 Logical Operators

Although logical operators are comparators the comparison is not always carried out. For example, if the left-hand object for an **and** gate is '0' (FALSE) the result will always be '0' and so the comparison need not be made. The operator is said to be "short-circuited" under such conditions. This means that the comparison is performed only if the value of the object on the left side of the operator is not sufficient to determine the result of the operation. Similar "short-circuits" occur for **or, nand** and **nor.**

The objects used on either side of binary logical operators must be of the same type. An object of type BIT cannot be compared with an object of type BOOLEAN.

5.4.2 Overloading Logical Operators

An architectural solution for simple gate functionality can be written using logical operators. For example; in the xor_gate of the example in Illumination Box 5.3.5 the architecture could have been written in the form:

```
architecture logic of xor_gate is
begin
    c <= a xor b;
end logic;
```

If *a* and *b* are input ports of type BIT the above architecture will model the functionality of an xor_gate.

The logical operator **xor** produces the same waveform value for assignment to *c* as would the following conditional assignment statement:

```
c <=      '0' when a = '0' and b = '0' else
          '1' when a = '0' and b = '1' else
          '1' when a = '1' and b = '0' else
          '0' when a = '1' and b = '1' else
          '0';
```

This is true *provided* the input ports are of type BIT. However, what would happen if the input ports were of, say, type Std_logic?

We have seen that logical operators are defined only for types BIT and BOOLEAN. We have also been made aware that the type of an object determines the values it can assume and the operations which can be performed on the object. Therefore, if the input ports a and b were to be defined as being of type Std_logic then the logical operator **xor** would, somehow, have to be defined for Std_logic type objects. The process of declaring other meanings for VHDL operators is called "overloading" and allows logical operators to be defined for objects other than BIT and BOOLEAN.

5.4.3 xor_gate

- **An architectural solution for simple gate functions can be written using logical operators**

- **Example**

```
architecture logic of xor_gate is
begin
        c <= a xor b ;
```

xor_gate Model

```
Library IEEE ;
use IEEE.Std_logic_1164.all ;

entity xor_gate is
        port (a, b : in Std_logic
                    c : out Std_logic ) ;
end xor_gate ;

architecture logic of xor_gate is
begin
        c <= a xor b;
end logic ;
```

5.4.3 xor_gate

If the objects a and b in the architecture given below are type Std_logic signals the logical operator **xor** must be defined for objects of this type:

```
architecture logic of xor_gate is
begin
   c <= a xor b;
end logic;
```

5.4.4 Overloading

When a standard operator is defined for a user defined object type the operator is said to be "overloaded" where:

1. Overloading is the process of defining how standard operators are applied to user declared types.
2. Overloading is part of the package in which the user declared types are declared.
3. The manner in which logical operators are applied to, say, 'Std_logic' objects is defined in a package called Std_logic_1164 (see Sections 5.3.4 and 5.3.5).

The following table shows how the logical operator **xor** is overloaded for application to Std_logic objects:

right/left	X	0	1	Z
X	X	X	X	X
0	X	0	1	X
1	X	1	0	X
Z	X	X	X	X

Table 5.1

NB. The code for the above would be contained in a subprogam called a **function** (See Chapter 11).

This is equivalent to the following conditional signal assignment statement:

```
c <=    '0' when a = '0' and b = '0' else
        '1' when a = '0' and b = '1' else
        '1' when a = '1' and b = '0' else
        '0' when a = '1' and b = '1' else
        'X';
```

Note that the last assignment in the list has changed with respect to the previous version in that the last '0' has become an 'X' (see paragraph 5.3.5).

Finally the xor_gate model which was the subject of the example in Section 5.3.5 may be rewritten as shown in the Illumination Box 5.4.3.

5.4.4 Overloading

- definition of how standard operators are applied to user defined data types

- Standard operators are not defined for user defined enumerated types

- Usually part of package in which data type is defined

- Example

  Manner in which logical operators are applied to Std_logic objects as defined in package Std_logic_1164

```
function "xor" (l, r : Std_logic)
return Std_logic

--  r\l  X  0  1  Z
--  X    X  X  X  X
--  0    X  0  1  X
--  1    X  1  0  X
--  Z    X  X  X  X
```

5.5 Delay

Thus far the models created in this text have ignored the question of delay (defined as the time period between cause and effect in a process) on the assumption that action and reaction were instantaneous. However, in the modelling of real systems, this clearly cannot be the case. Delay must be taken into account and it may be shown that there are two major kinds of delay which must be addressed within VHDL. Also, the fact that "several processes can happen at the same time", or concurrently, is a feature of modelling which is not present in 'von Neumann-type' programming languages but which must be catered for within a hardware modelling language.

The two major topics concerning delay in VHDL are, therefore:

1. Delay selection

   Delay selection is concerned with how delay interacts with signals. In VHDL there are two algorithms for selecting the interaction between delay and signal transfer. These are:

   · Inertial

   Inertia is resistance to change. In an inertial system a change takes place only if the force causing the change is applied for a time sufficient to overcome the inertia.

   · Transport

   'To transport' is to carry from one point to another. In a transport system a change takes place independently of the duration of the force causing the change.

2. Internal delay

   VHDL is a *hardware* modelling language. Unlike software programming languages, it is designed to model hardware processes. Hardware has the potential for concurrent execution, i.e. several processes can take place at the same time. The mechanism for implementing concurrent execution is tied to a construct in VHDL called 'delta delay' and is referred to as internal delay.

5.5 Delay

- Time period between cause and effect

- Delay selection:
  - Inertial
  - Transport

- Internal delay
  - Delta

Inertial delay

- Characteristic of any system where there is a time lag between cause and effect

Example

- undelayed model:
  c <= a and b
- delayed model:
  c<= a and b after one_time_period

Time Period	a	b	c
0	0	0	0
1	1	1	1
2	0	1	0
3	1	1	1
4	1	0	0

Time Period	a	b	c
0	0	0	0
1	1	1	0
2	0	1	1
3	1	1	0
4	1	0	1

5.5.1 Inertial Delay

Inertial delay is characteristic of any system in which there is a time lag between cause and effect. In an un-delayed model the effect of a signal change would be seen immediately. This response is what we have seen in the models developed so far in this text and is sometimes referred to as 'zero delay'. When forces are applied to input ports the result is seen immediately on the output port. This type of response is unrealistic in the physical world since, no matter how fast the response to stimulus, there will always be some lag between application and response. In electronic systems this lag may be measured in units as small as femtoseconds, but it will exist nevertheless.

To model inertial delay requires the addition of an **after** clause to signal assignment statements. This clause specifies the amount of delay which is to occur before the effect appears.

Delay is measured in units of time, therefore the time period in an **after** clause must be expressed in time units consistent with the predefined type TIME. Time units in this type extend from femtoseconds (fs) to minutes (min). Normally the magnitude of the time period is expressed in whole number units; however, magnitude may be expressed in fractional units if it can be resolved into a whole number of base units (in this case - femtoseconds).

5.5.2 The after Clause

Inertial delay is modelled with the **after** clause. An **after** clause may be placed after the signal assignment expression and before the semicolon in any signal assignment statement, for example:

1. a_signal <= b_signal **after** 10 ns;

   In this statement signal *b_signal* is assigned to *a_signal* after a delay of 10 ns. Or, alternatively speaking, whenever *b_signal* changes *a_signal* will assume the new waveform 10 ns later. Note that this change will only occur *provided the duration of the change is equal to, or greater than, 10 ns.*

2. d_out <= a **nand** b **after** delay_time;

   In this statement the logical operation **nand** is applied to the signals *a* and *b*. The result of the **nand** operation will be assigned to signal *d_out* after the time specified in the object 'delay_time'. The object 'delay_time' must be a type TIME object.

3. out_put <= '1' **after** 10 ns,
'0' **after** 20 ns,
'1' **after** 30 ns,
'0' **after** 40 ns;

5.5.2 After Clause

- **Inertial delay is implicitly modelled with the after clause**
- **Format**

  new_signal <= signal_expression **after** time ;
- **Examples**
  - ```
    a_signal <= b_signal after 10 ns ;
    d_out  <= a nand b after delay_time ;
    out_put <= '1' after 10 ns,
                   '0' after 20 ns,
                   '1' after 30 ns,
                   '0' after 40 ns;
    q <= '1' after 10 ns when b = '1' else
         '0' after 5 ns ;
    ```

5.5.3 A Delayed xor_gate Model

```
library IEEE ;
use IEEE.Std_logic_1164.all ;

entity xor_gate is
    port (a, b : in Std_logic ;
             c :  out Std_logic) ;
end xor_gate ;

architecture logic of xor_gate is
begin
    c <= a xor b after 10 ns ;
end logic ;
```
```

In this statement a signal driver for the signal out_put is established. This statement is very similar to the static signal assignment discussed at the beginning of this chapter. It differs in that a series of waveforms, rather than a single waveform, is assigned to the signal.. This creates a container, of projected output waveform values and application times, called a driver. A projected output waveform consists of a sequence of one or more transactions, where each transaction is a pair consisting of a value component and a time component. For a given transaction the value component represents a value that the driver of the signal is to assume at some point in time, and the component specifies which point in time. These transactions are ordered with respect to their time components. A driver always contains at least one transaction[1]. Multiple drivers will be discussed later in the book (Chapter 13, Section 13.3).

4. q <= '1' **after** 10 ns **when** b = '1' **else**
'0' **after** 5 ns;

This statement shows how the **after** clause can be applied to the conditional signal assignment statement. Note that the **after** clause precedes the **when** clause.

To summarise; the **after** clause:

a) Allows inertial delay to be added to a model
b) Appears in signal assignment statements after the determination of the waveform value
c) Must be followed by specific time units or by an object of type TIME.

5.5.3 A Delayed xor_gate Model

The model shown in Illumination Box 5.5.3 is a slightly modified version of the model appearing in the Illumination Box 5.4.3. The only change is the addition of an **after** clause to the signal assignment statement.

If this model were to be simulated, signal *c* would be assigned the **xor**ed value of signals *a* and *b* ten ns after a change in either *a* or *b*.

It is possible, however, that the waveform of signal *c* would not change even though it had been assigned a new value. Suppose that signals *a* and *b* were '1' and '0' respectively and were then changed to '0' and '1'. Ten ns after the change *c* would be assigned a value '1'. Since *c* is '1' both before and after the assignment, there is no change in value. This type of action is called an 'activity'. If the value of *c* had changed, in addition to an activity, there would have been an 'event'. The terms 'activity' and 'event' will appear in discussion throughout this text.

Consider another possible set of changes. Suppose signals *a*, *b* and *c* were '1', '0' and '1' respectively, Signal *a* then changes to '0', but 5 ns after it changed to '0' it changes back to '1'. What will happen to signal *c*?

5.5.4 Signal Duration

- **Signals whose durations are less than the delay specified in the after clause are ignored**
- **Example**

signal assignment statement

a <= b after 50 ns ;

Stimulus set

force	b	0	0
force	b	1	50
force	b	0	100
force	b	X.	150
force	b	0	175
run	250		

Result

25 ns unknown pulse on b at 150 ns will not be passed to a.

5.5.4 Signal Duration

Signal waveforms whose durations are less than the time delay specified in the **after** clause are ignored.

Consider the example shown below;

The signal assignment statement:

a <= b **after** 50 ns;

assigns the waveform of signal b to signal a fifty ns after the waveform on signal b has an activity.

Now consider the following stimulus set:

force	b	0	0
force	b	1	50
force	b	0	100
force	b	X	150
force	b	0	175
run	250		

where the first column indicates the stimulus to be activated in b at the times, in nanoseconds, given in column two (assuming the simulator default unit of time to be 1 ns).

This stimulus set produces a 25 ns 'unknown' pulse between 150 and 175 ns; but the delay in the signal assignment statement is 50 ns. Therefore it is to be expected that the 25 ns pulse would be ignored.

A test with a simulator might produce a trace which looks like the following:

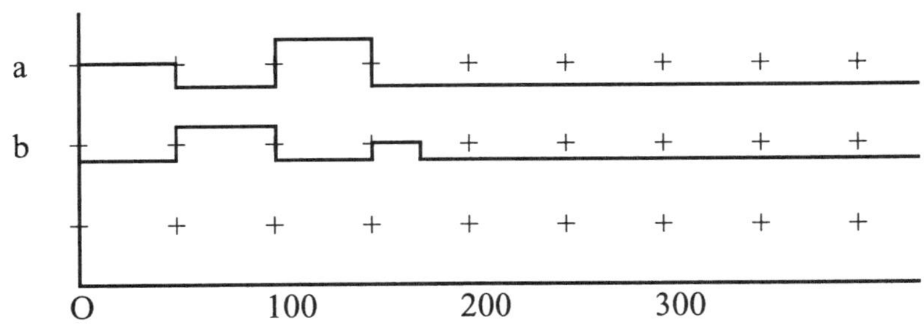

Figure 5.1 Inertial Delay

On the assumption that X (indeterminate) is traced along the + signs and that bit '1' is above the line and bit '0' below, note that the 25 ns pulse on signal b is not transferred to signal a. Also, a is indeterminate for the first 50 ns after the start of the run period. To be transferred, the waveform must exist for at least as long

5.5.5 Transport Option

- **Passes signals regardless of duration**
- **Used when signals are delayed but 'filtering' effect is unwanted**
- **Example**

signal assignment statement

a <= **transport** b **after** 50 ns ;

Stimulus set

force	b	0	0
force	b	1	50
force	b	0	100
force	b	X	150
force	b	0	175
run	250		

Result

25 ns unknown pulse on b at 150 ns will be passed to a.

as the amount of time in the delay. In the above case, a signal having less than 50 ns delay with respect to the signal immediately preceding it will be ignored. Note that a signal assignment statement with an **after** clause of 0 time is equivalent to a signal assignment statement with no **after** clause.

5.5.5 Transport Option

The 'filtering' effect produced by the **after** clause is one approach to the modelling of physical reality, However, there are instances when the modeller might wish to pass signals regardless of duration. This is not a problem if the length of the time delay is zero; all waveforms, regardless of duration, are passed under these conditions. The problem occurs when the modeller wishes to pass all waveforms regardless of duration but with a specified delay greater than zero. Consider the example in Illumination Box 5.5.5. The signal assignment statement

a <= **transport** b **after** 50 ns;

assigns the waveform of signal b to signal a fifty ns after the waveform on signal b has an activity, regardless of the duration of signal b.

Now consider the following stimulus set

```
force    b    0    0
force    b    1    50
force    b    0    100
force    b    X    150
force    b    0    175
run      250
```

If this signal assignment statement were identical to the signal assignment statement in Section 5.5.4 we would not expect the 'unknown' pulse to appear on the output signal. The delay in the signal assignment statement is 50 ns. The duration of this pulse is 25 ns. Therefore we would expect this pulse to be ignored. However, this is not the case since the **transport** option passes all waveforms regardless of duration.

A test with a simulator might produce a trace which looks like the following:

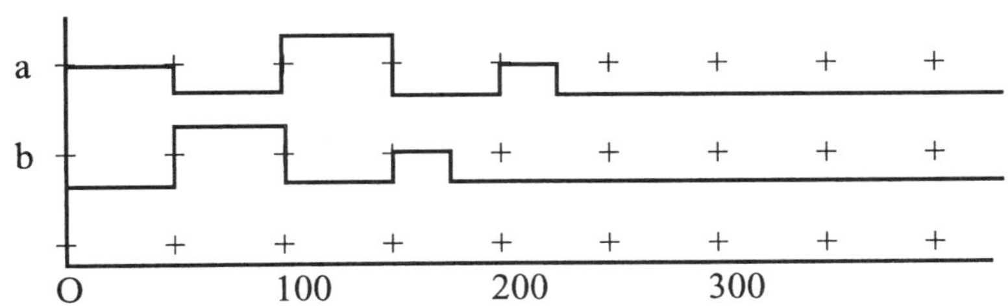

Figure 5.2 Transport Delay

The signal waveform is passed with 50 ns delay regardless of its duration.

Reference

1. *IEEE Standard VHDL Language Reference Manual (ANSI/IEEE Std 1076-1993)*

5.6 Exercises

5.6.1 What is wrong, if anything, with the following description of a 2-input or_gate?

```
entity or_gate is
    port ( a, b : in Std_logic ;
           c : out Std_logic) ;
end or_gate ;

architecture data_flow of or_gate is
begin
    c <=    '1' when a = '1' or b = '1' else
            '0' ;
end data_flow ;
```

5.6.2 Write the behavioural architectural description of a two input nor_gate having inertial delay of 10 ns.

5.6.3 a) When is the **transport** option useful?
b) When is the **transport** option not useful?

5.6.4 Write a behavioural description for a 1-bit comparator in which a > b.

5.6.5 List, and define, the nine values for the Package IEEE.Std_Logic_1164

5.6.6 Write a structural description for a 4-bit comparator in which a = b.

5.6.7 Create a model for an entity which receives two signals and which generates one output signal. If both inputs are '1' the output is to be '0'. Any combination of '0' with another of the states ('X', '0', '1', 'Z') should produce an output of '1'. If either or both inputs are 'Z' or'X' and neither of the inputs is '0' the output is to be 'X'. All outputs are to be delayed 25 ns from the change of the input.
If '0' appears on either of the inputs the output is to be '1'(for example, a '0' and 'X' produces a '1', as does 'Z' and '0'). However, a '1' and an 'X' or a 'Z' produces an 'X'.

5.6.8 The following model contains a number of VHDL syntax errors. What are they?

```
entity 2_to_1_encoder is        -- NB. See page 472 w.r.t. extended
    port (a_in, b_in : in BIT ;                 -- identifiers
                out : out BIT ) ;
architecture function of 2_to_1_encoder is


end function ;
```

5.6.9 Overload the logical operator **nand** for application to Std_logic objects.

Chapter 6

Multiple Architectures and Concurrency

Overview

This chapter, like Chapter 5, builds a deeper understanding of concepts which are already partly understood, opens discussion on topics which have been introduced but not explained, and introduces new concepts. Many of the topics introduced in this chapter are interrelated. For example, concurrency is related to delta delay which, in turn, forces the issue of iterations and the necessity of having some way to trace execution.

In this chapter you will learn

- That there may be more than one architecture per entity per declaration
- The limitations imposed by signal drivers
- The meaning of concurrent execution and some of its effects
- The purpose of Delta Delay and how it affects model execution
- The concept of iterations

6.1 Multiple Architectures

1. Association is done at the time of compilation
2. The entity must be compiled *before* any of the architectures associated with it are compiled
3. There can be only one entity (multiple entities require multiple components)
4. Entities and architectures may exist in the same source file, or in different source files, provided the entity is compiled first
5. If an entity is recompiled, all architectures associated with it must also be recompiled. On the other hand, an entity need not be recompiled when an associated architecture is recompiled

6.1 Multiple Architectures

It has been stated (Section 4.5.2) that multiple VHDL architectures may be associated with a single entity. This generates several questions. One is "How is this done?" and another is "Which architecture should be used during simulation?".

Associating multiple architectures with a single entity is very easily done. Such association is made via the entity name which appears in the architecture declaration, therefore any architecture within a component that has the entity name in its declaration is associated with the entity declaration for that component when it (the architecture) is compiled. This implies several aspects of architecture/entity association:

1. Association is done at the time of compilation.
2. The entity must be compiled *before* any of the architectures associated with it are compiled.
3. There can only be one entity (multiple entities require multiple components).
4. Entities and architectures may exist in the same source file, or in different source files, provided the entity is compiled first.
5. If an entity is recompiled, all architectures associated with it must also be recompiled. On the other hand an entity need not be recompiled when an associated architecture is recompiled.

These aspects of multiple architectures shape the way model development may proceed. For example, it is probably better to place the entity and the architecture in separate files if it is anticipated that there will be frequent changes to the architecture. This will prevent having to recompile the entity every time the architecture is recompiled. It should be recognised, however, that recompiling an entity using this approach could create its own problems in so far as *all* associated architectures must be recompiled and, if these exist in separate files, one, or more, may be overlooked.

The question remains as to which of the multiple architectures should be associated with the entity during simulation. It should be obvious that only one architecture may be used for each instance of the design during simulation since if more than one architecture were declared the simulator would be faced with conflicting instructions.

The architecture to be used during the simulation is identified by the model property on the symbol or in the 'viewpoint' for the design. If a symbol and viewpoint are not specified for a design when using an HDL CAE tool, the tool should, normally, generate a default model property.

6.1.1 Multiple Architectures Example

```
entity multiplexer is
      port( sel, a, b : in BIT ;
                  d_out : out BIT) ;
end multiplexer ;

architecture behaviour of multiplexer is
begin
process (sel)
begin
      if sel = '1' then
            d_out <= a else b ;
end behaviour ;

architecture RTL of multiplexer is
begin
   d_out <= (sel and a) or (not sel and b) ;
end RTL ;
```

6.1.1 Multiple Architectures Example

The following example depicts possible architectures for a 2-input multiplexer (NB. The description given below is for illustration only and is therefore incomplete. See Chapter 3, Section 3.3 for a complete example):

```
entity multiplexer is
    port (sel, a, b : in BIT ;
            d_out : out BIT ) ;
end multiplexer ;
architecture behaviour of multiplexer is
begin
process (sel) -- see Chapter 7 (Sequential Statements and
begin                                          -- Processes)
    if sel = '1' then
            d_out <= a else b ;
end behaviour ;

architecture RTL of multiplexer is
begin
    d_out <= (sel and a) or (not sel and b) ;
end RTL ;
for application : multiplexer
    use entity multiplexer (behaviour) ;
```

NB: name in parentheses tells system to use the 'behaviour' version architecture of a component called multiplexer. This process will be dealt with in detail in Chapters 14 and 16.

6.2 Signal Drivers

- **Containers for projected waveforms of signals**
- **Created when a signal assignment is made**
- **Example**

```
--      filling a driver
        clock <=    '0'
                    '1' after 5 ns,
                    '0' after 10 ns,
                    '1' after 15 ns;
```

clock driver

time	signal
0	0
5	1
10	0
15	1

6.2 Signal Drivers

The concept of a 'signal driver' was introduced in Chapter 5 (Section 5.1) but was not discussed beyond saying that a signal driver was a container for projected waveforms of a signal.

It is important to realise that a signal driver is created by a concurrent VHDL statement in which a waveform is assigned to a signal object. IN the example shown in Illumination Box 6.2 the signal assignment statement

```
clock  <=  '0',
           '1' after  5 ns,
           '0' after 10 ns,
           '1' after 15 ns;
```

creates and fills a driver with the indicated waveforms and associated times. NB: Statements of this form are called "static assignments", they are executed once only during the simulation of the model. This is normally done when a simulator such as Quicksim® is first invoked on the model. Therefore, at any time in the simulation beyond 15 ns the value of the object *clock* will be '1'. The usefulness of such statements is, therefore, limited; however, the concept is very important.

Notice that the signal driver is said to contain "projected waveforms" and not "future waveforms". The difference between "projected" and "future" is that "projected" implies some possibility that the value will not occur while "future" means that the value will definitely occur. Waveforms in drivers are *projected.*

They indicate what might occur and the time at which it might occur based upon what has already happened; however, future events may change what will happen.

While the concept of what a driver is may be relatively simple (a container of projected waveforms), there are certain rules regarding signal drivers.

1. Multiple drivers can be created for the same signal object if waveforms are assigned to the object in multiple concurrent statements.
   The creation of multiple drivers for a single signal creates a conflict as to what is the correct value. This problem can be resolved using "resolution functions". However, since resolution functions are not discussed until Chapter 13 it is important to realise that, for now, waveforms should not be assigned to the same signal in different concurrent statements.
2. The projected waveforms in the driver can be changed before they are applied to the signal. This means that there must be some "rules" for establishing how values are updated in the driver.

These topics will now be discussed in more detail.

6.2.1 Filling Drivers

- **Negative time is not allowed**
- **Assignments within a single signal assignment statement must be made in ascending time order**
  Example

```
clock <=    '0',
            '1' after 5 ns,
            '0' after 10 ns,
            '1' after 15 ns;
```

- **Assignments to a single signal by multiple concurrent statements create multiple drivers which must be "resolved"**
  Example

```
signal_out   <=   signal_in_1 after 10 ns
signal_out   <=   signal_in_2 after 10 ns
```

- **Multiple executions of assignment statement modify projected waveforms in driver**

6.2.1 Filling Drivers

There are certain rules about signal drivers which must be observed.

1. Negative time is not allowed.
   Drivers contain projected waveforms. All times associated with these waveforms must be in the future with regard to the current time. It is not possible to alter what has taken place in the past.
2. The assignment of waveforms and associated times within a single signal assignment statement must be made in ascending time order.
   The first example shown in Illumination Box 6.2.1 shows the correct way to create a signal assignment statement which assigns multiple waveforms and times. The following example is incorrect:
```
clock   <=   '0' after 10 ns,
             '1' after 15 ns,
             '0' after 20 ns,
             '0' after  5 ns;-- incorrect, times
                                not in ascending order
```
   The waveforms and associated times must be written in ascending time order within a single signal assignment statement.
3. Waveform assignments to a single object by multiple concurrent statements create multiple drivers.
```
signal_out   <=   signal_in_1 after 10 ns;
signal_out   <=   signal_in_2 after 10 ns;
```
   Each of the above statements creates a driver for *signal_out*. If both input signals have different waveforms, which waveform will be assigned to *signal_out*?
   Without some procedure for "resolving" which waveform takes precedence there is no way of determining the final waveform. Resolution functions are discussed in Chapter 13. Until resolution functions are presented models should have only one concurrent signal assignment for a given object.
4. Scheduling of events within a driver by a single signal assignment statement is based on several factors; for example:
   · Concurrent Signal Assignment
   If a single concurrent signal assignment statement with a delay (**after**) is executed multiple times, the driver of the signal object is modified each time the statement is executed. Without the **transport** option, all events in the future (and in the past) of the event being added are deleted except for the current value being driven onto the object. For past events (events that occur before the time associated with the value being placed in the driver) all values different from the event being

6.3 Concurrency

- Occurrence of two, or more, events in the same time period
- In the following system, a change in signal b may cause a concurrent change in all of the output signals

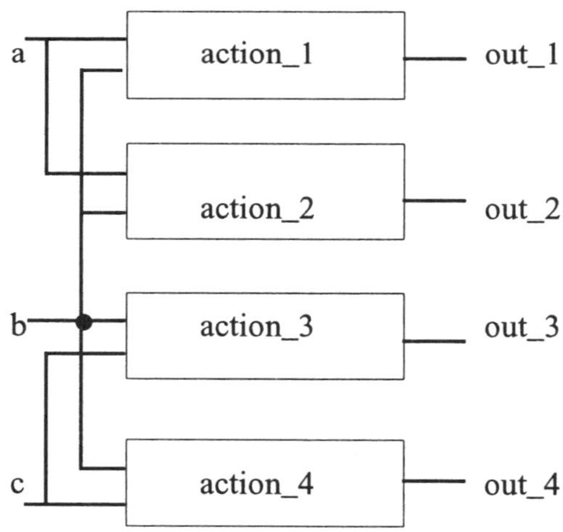

scheduled are deleted (those with the same value remain). With **transport**, all past events are retained but "future" events are deleted.
Sequential Signal Assignment
This will be discussed after sequential signal assignment is introduced.

6.3 Concurrency

Concurrency refers to the simultaneous occurrence of two or more events. Things that are concurrent are said to happen at the same time.

Electronic hardware provides many examples of concurrent events. Consider the system illustrated in Illumination Box 6.3. Three inputs (*a, b* and *c*) are connected to four devices (*action_1, action_2, action_3* and *action_4*). Each device produces one output (*out_1, out_2, out_3* and *out_4*). Input *b* is connected to all four devices. Input *a* is connected to *action_1* and to *action_2*. Input *c* is connected to *action_3* and to *action_4*.

A change in signal *b* in this system may cause a change in all of the output signals. Because the function of each of the devices is unknown, a concurrent change in output "may", instead of "does", occur. It is very possible that the delay in each of the devices is different, in which case each of the output signals would change at a different time. However, if we assume that when signal *b* changes the output signals all change 10 nanoseconds later, the devices will produce concurrent outcomes.

In the past modelling concurrent systems has been difficult. Many earlier modelling languages executed sequentially. Therefore, special techniques had to be devised in order to correctly model concurrent actions. VHDL has overcome this problem by creating a set of "concurrently" executed statements.

The existence of such statements is unsettling to some modellers. They are used to a sequential flow through the model code; but concurrent statements in VHDL may be executed in any order. These actions are concurrent so they must execute *at the same time.*

It should be evident that concurrent execution of statements is a major step forward in the creation of models which emulate actual systems. It may also be evident that concurrent execution will require much explanation and discussion. The next few sections of this chapter will attempt to explain some of the aspects of concurrency and concurrent execution.

6.3.1 Concurrent Execution

- **to model reality VHDL processes certain statements concurrently**
- **Example**

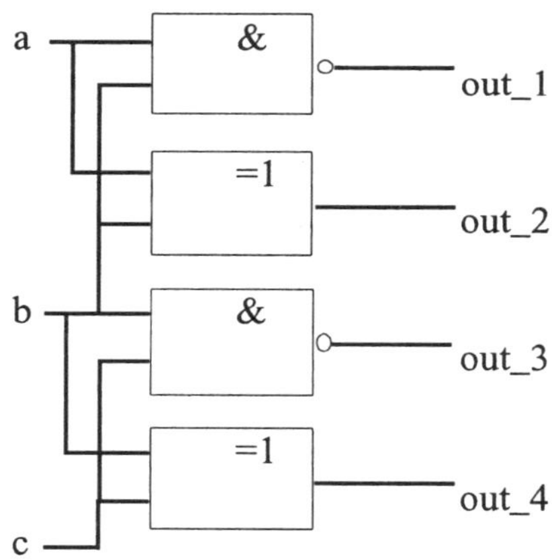


```
architecture example of concurrent is
begin
        out_1 <= a nand b;
        out_2 <= a  xor  b;
        out_3 <= c nand b;
        out_4 <= c  xor  b;
end example;
```

6.3.1 Concurrent Execution

It has been previously stated that VHDL has the ability to process certain statements concurrently. Since many actions in the real world take place at the same time this ability permits better models to be constructed.

Consider the example in Illumination Box 6.3.1. This system shows the connection of signals to two nand_gates and to two xor_gates. Except for the substitution of gate symbols for block systems, this drawing is identical to that in Illumination Box 6.3.

How could this system be modelled? Consider the following architecture:

```
architecture example of concurrent is
begin
    out_1 <=    a NAND b;
    out_2 <=    a  XOR b;
    out_3 <=    c NAND b;
    out_4 <=    c  XOR b;
end example;
```

If all gate action is assumed to have zero delay, this architecture correctly models the action of the system.

The immediate, conventional, response might be "Of course! The signal a is NANDed with signal b and the result assigned to signal *out_1*. Then the signal a is XORed with the signal b and the result is assigned to signal *out_2*; and so on".

This would be a wrong assessment of the situation being modelled. If the evaluation were to be conducted as described within the quotation marks above, the model would not, accurately, be depicting the way the system acts. When signal b changes, all of the output signals can, potentially, change too; and if they did, all would change at the same time. What is described in the previous paragraph is the conventional software programmer's interpretation that one output signal changes, then another, then another and so on. This is not how the system functions and not how it has been described using VHDL. They may all, potentially, change at the same time.

VHDL does, in fact, change all of the output signals at the same time (NB. "activity"; see Section 6.4). This statement invites a number of questions since computers process statements sequentially. Whilst it is true that statements are executed sequentially, VHDL has a mechanism for determining which statements should be executed concurrently, executing these statements and applying all the results at the same time. Hence, it makes no difference which statement is executed first by the computer.

6.3.2 Concurrent Statements

- **The architecture of a model is made up of one, or more, concurrent statements**

- **Concurrent statements**

    - **block** statement
    - **process** statement
    - **assertion** statement
    - signal assignment statements
    - procedure calls
    - component instantiations

- **Each concurrent statement represents a unit of functionality**

6.3.2 Concurrent Statements

The architecture of a model is made up of one or more concurrent statements (actually, a "null" architecture, i.e. one containing no statements, can exist but such an architecture has no functionality).

There are six concurrent statement types in VHDL:

1. **block** Statements
2. **process** Statements
3. **assert**ion Statements
4. **signal** Assignment Statements
5. **procedure** Calls
6. **component** Instantiation Statements

Each concurrent statement represents a unit of functionality. In some cases the concurrent statement simply encloses the functionality; in other cases it is the actual functionality. In all cases concurrent statements have the potential of being executed at the same time as one or more other concurrent statements. This is indicated by the name "concurrent" statement.

In addition to concurrent statements, there is a class of statements called sequential statements. These statements will be introduced and discussed later in this volume. However, it is important to note that sequential statements can only appear within concurrent statement control. Architectures are made up of concurrent statements which may, or may not, contain sequential statements.

The following architecture is made up of four concurrent signal assignment statements. We have already learned to call this type of concurrent statement a signal assignment statement.

```
architecture example of concurrent is
begin
    out_1 <=    a NAND b;
    out_2 <=    a  XOR b;
    out_3 <=    c NAND b;
    out_4 <=    c  XOR b;
end example;
```

We have heard that concurrent statements do not execute in a sequential manner. If this is so, what is it that determines which statements execute and when they execute? The answers to these questions are the subject of the next section.

6.4 Statement activation

- **Signals connect concurrent statements**
- **Concurrent statements activate when there is an event on a signal "entering" the statement**

- **Example**

```
architecture example of concurrent is
        signal c : Std_logic;
begin
        c       <= a nand b;   -- nand gate
        out_1   <= c xor b;    -- xor gate
end example;
```

  - An event on signal a activates the nand_gate statement
  - An event on signal b activates both the nand_gate and the xor_gate statements
  - An event on signal c activates the xor_gate statement

6.4 Statement Activation

Before we continue, let us consider this somewhat simplistic view of electronic systems:

Components represent blocks of functionality. These blocks are connected by wires carrying signals. The functionality of the block establishes the relationship between the input to the block and the output from the block. When the input to the block changes, the functionality determines what changes have to be made in the output and makes those changes.

Now think about this view of VHDL:

Concurrent statements represent blocks of functionality. These blocks are connected by signals having specific waveforms. The functionality of the concurrent statement establishes the relationship between the input to the block and the output from the statement. When the input to the statement changes , the statement is activated so that the functionality can determine what needs to be done.

The parallels are not coincidental. VHDL is designed to parallel reality. Blocks of functionality in reality can be components (cards, chips, devices or entire systems); blocks of functionality in VHDL are concurrent statements.

Wires in reality carry signals; signals in VHDL represent conduits for carrying values from one concurrent statement to another. Components act on new inputs to produce new outputs; concurrent statements activate when signals, inputted to them, change.

Hence:

1. Signals connect concurrent statements
2. Concurrent statements activate when there is an event on a signal involved in the process which creates the output. Concurrent signal assignment statements are said to be "sensitive" to signals to the right of the assignment operator (<=).

Examine the example of Illumination Box 6.4. The schematic shows a "real" system consisting of the interconnection of a nand_gate and an xor_gate. The statement c <= a NAND b; represents the nand_gate. The statement out_1 <= c XOR b; represents the xor_gate. The signal c connects the output of the nand_gate with one of the inputs of the xor_gate.

The concurrent statement will activate when:

- An event on signal *b* causes both statements to activate. The signal *b* appears on the right-hand side of the assignment symbol; therefore, both statements are "sensitive" to changes in *b*.
- An event on signal *a* causes the nand_gate statement to activate.
- An event on signal *c* causes the xor_gate statement to activate.

6.5 Defining additional signals

- **Signals may be declared within declarative area of architecture**

- **Format**

```
signal signal_name : type;
```

- **Example**

```
architecture example  of extra_signal is
    signal int : Std_logic;
begin
    int     <= a and b;
    out_l <= c xor int;
end   example;
```

6.5 Defining Additional Signals

During the discussion on Illumination Box 6.3.1, VHDL signals were represented as having the same function as wires. Signals are conduits for waveforms from one block of functionality to another block of functionality. For the moment, this means that signals conduct waveform values to and from concurrent signal assignment statements.

Objects declared in the **port** clause of the entity declaration are signals. They conduct waveform values in and out of the model. In the simple models constructed in the first chapters of this volume there is, usually, only one block of functionality, hence the port signals were sufficient. Values conducted into the model established what values were output from the model.

In this chapter we have seen that the architecture may consist of several blocks of functionality. In some cases, the output from one block of functionality is input to another block of functionality. The signals declared in the **port** clause are not the correct medium for this type of transfer. They are primarily for establishing the conduits between the model and its environment. Internal signals are declared separately.

Signals may be declared within the declarative area of the architecture. This is the area immediately between **architecture** and **begin**. Signal declarations are formatted as follows:

signal signal_name : **type** ;

Notice that no mode (**in, out, inout**) is given in the declaration of internal signals. All internal signals are of mode **inout**. This is consistent with the purpose of internal signals (to conduct values from one concurrent statement to another).

The mode must be **out** from the sending block and **in** from the receiving block. The mode **inout** serves this purpose.

In the example in Illumination Box 6.5, an internal signal is needed to conduct the output of the and_gate to the xor_gate. To do this, a signal named *int* is declared in the declarative region. The signal object *int* is declared to be type *Std_logic*. The type must match the set of values which might be assigned to the signal object. In this case, if signals *a* and *b* are *Std_logic* signals, *int* must also be *Std_logic*.

To conclude, it should be remembered that:

1. Internal signals are declared in the declarative area of the architecture.
2. All internal signals are of mode **inout**.
3. The type of the internal signal must correspond to the value set which the signal will carry (the type defines the values the object may assume).
4. The initial value of an internal signal may be stated or defaulted. If defaulted, it is the left-most value in the value set of the type.

6.6 Concurrent signal change

- Signals which change during concurrent execution are "queued" to change in the future
- Signal assignment statements without an explicit after clause are given a delay of one delta
- Example

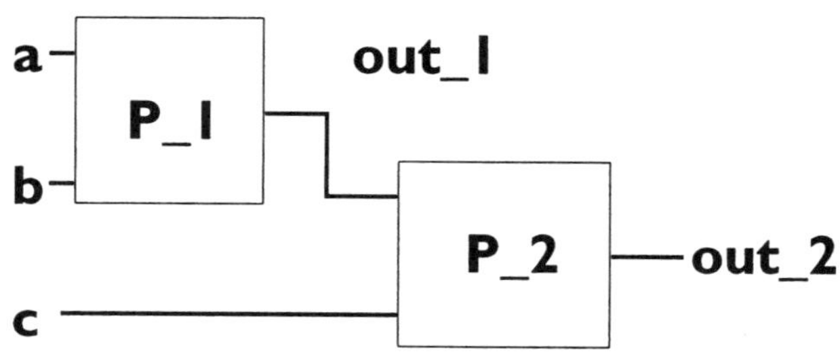

6.6 Concurrent Signal Change

We have learned that certain statements in VHDL can execute at the same time. These statements are called concurrent statements. The concurrent statement which we have used is the concurrent signal assignment statement. The question which has not been answered yet is; "How can several activated concurrent signal assignment statements make simultaneous assignments when each statement must be evaluated individually?".

Most computers process statements sequentially. This is evident when observing the operation of single stepping, e.g. the trace mode on the MC68000. Activated statements are obviously evaluated one at a time. It would seem that the assignment from the statement which was evaluated first would be made before the statement which is evaluated last.

Assignment of waveforms to signals is not made at the same time the waveform is determined. When several statements are activated at the same time, signal waveforms which change during the concurrent execution are "queued" to change in the future. In the case of concurrent signal assignment statements which have an associate delay, the change is queued to occur at a specified interval from the 'present'. However, concurrent signal assignment statements can be written with no explicit delay (no **after** clause). In these cases a delay of one delta time is given to the assignment.

Another way to look at this process is as follows: As a time period is entered the system checks to see whether any stimuli are to be applied during this period. If there are stimuli pending for the period, the system determines which statements are activated by those stimuli. Once all of the statements which activate are identified, the system evaluates each statement and retains the results of the evaluations. When all statements activated by the initial stimuli have been evaluated, the results are assigned.

Although the statements are not evaluated concurrently, the result is the same as if they were.

Each iteration is the evaluation of all of the activated statements. At the end of each iteration the values are assigned and the system checks to see if any 'new' statements are activated owing to the results just assigned.

6.7 Delta Time

The VHDL mechanism for "concurrent" evaluation of statements is dependent upon the construct called "delta time".

Delta time is an infinitely small amount of time used when emulating concurrent evaluations. It is so small that an infinite number of delta times adds up to zero.

Delta time does not advance simulation time.

6.7 Delta Time

- **A small amount of time used in concurrent evaluations**
- **An infinite number of delta times add up to zero time**
- **All activated statements are evaluated and the results assigned one delta time later**
- **Example**

out_l **<=** a **nand** b

time	a	b	out_l
0	0	1	1
1	1	1	1
1 + ^	1	1	0

At the beginning of a time period the system decides what statements, if any, are activated by the change of a signal in their "sensitivity list". These statements are then evaluated and the results of these evaluations assigned one delta time later.

The example shown in Illumination Box 6.7 demonstrates the effect of delta time. The statement above the table represents the architecture of a model. The signal waveforms at time zero are stable with *a* equal to '0', *b* equal to '1' and out_1 equal to '1'. Simulation time is advanced one time period (to '1' in the table) and signal *a* changes to '1'. This signal change activates the statement. The statement is evaluated, but the result of the change is not assigned until one delta time later ('1' plus delta in the table).

After all activated statements have been evaluated and the assignment of the results of these evaluations made, the system checks to see if there are any subsequent activations based upon the signal changes just made. If subsequent activations occur, another round of evaluations, assignments and checks are made. Each round in this procedure is called an "iteration".

6.8 Iterations

The term "iteration" used here refers to the subsequent activation of statements caused by signal changes produced during the execution of previously activated statements. Each iteration takes place in one delta time. For most purposes an iteration and a delta time can be thought of as the same thing.

By convention, the first iteration is composed of those statements activated by the initial stimuli (forces). Statements activated by the signal changes produced during the execution of the first iteration are referred to as the second iteration. Subsequent activations are referred in ascending order of number.

The example depicted in Illumination Box 6.8 shows a modelling situation where two iterations occur.

1. At time zero the model is in a stable state situation. Signal *a* is '0', signal *b* is '1', signal *c* is '1', signal *out_1* is '1' and signal *out_2* is '0'.
2. Time is advanced into the first time period and signal *a* is changed to '1'. This change activates the statement *out_1* <= *a* **nand** *b*.
3. The statement is evaluated and one delta time later the value of the signal *out_1* is scheduled to change to '0'.
4. The system then checks and finds that this change causes the statement *out_2* <= *out_1* **xor** *c* ;
5. This statement is evaluated and one delta time later signal *out_2* is changed to '1'.
6. The system then checks and finds that no statements are activated by this change.

The number of iterations which occur depends upon the structure of the model.

6.8 Iterations

- **Subsequent activation of statements caused by signal changes produced during execution of previously activated statements**

- **Example**

```
architecture example of  iterations is
        signal out_1 : Std_logic ;
begin
        out_1 <= a nand b ;
        out_2 <= out_1 xor c ;
end example ;
```

time	a	b	c	out_1	out_2
0	0	1	1	1	0
1	1	1	1	1	0
1 + ^	1	1	1	0	0
1 + 2^	1	1	1	0	1

It is possible to create a model which infinitely iterates. If this were to happen, the simulation of the model could not proceed.

To prevent a simulation from being trapped in an infinitely iterating structure commercial simulators have limits imposed upon the number of iterations which can occur. These can normally be set by the user but if not so set a default limit will be imposed.

6.9 Exercises

6.9.1. Write a description for a 1-bit comparator.

6.9.2 The following code was written to generate a clock with an asymmetrical mark-space ratio. It would not compile. Why?

```
clk <=      '0' after 5 ns ;
            '1' after 10 ns ;
            '0' after 5 ns ;
            '1' after 10 ns ;
            '0' after 5 ns ;
                    •
                    •
                    •
```

6.9.3 Write
   a) Structural code and
   b) Behavioural code

for the sum and carry model of a 1-bit full adder with a carry input.

6.9.4 Draw timing diagrams which illustrate the following transactions on signals *b* and *d*.

```
architecture combining of signals is
signal a, b, c, d, e : BIT := '1' ;
begin
    b <= a nand c ;
    d <= b xor e ;
end combining ;
```

6.9.5 Write a description of a model having inputs a, b, c and output z which requires the following concurrent outcomes.

```
k = a and not b
l = not a and b
m = k or l
r = c and not m
s = not c and m
z = r or s
```

Draw a diagram which depicts the transactions necessary to fulfil the outcome for z. Can you suggest a simpler model for the circuit? Write the architecture for this model and add this multiple to the previous description.

6.9.6 Redesign the one bit comparator of Exercise 6.9.1 so that several may used in combination to form a multiple bit comparator.

Chapter 7

Sequential Statements and Processes

Overview

In this chapter a second concurrent statement type, the process statement, is introduced. It differs from the concurrent signal assignment statement in that it is a container of statements. Statements within a process statement are executed sequentially, hence the description - sequential statements.

In this chapter you will learn

- The concepts of sequential operations
- The concepts of sequential statements
- The concept of the variable
- The difference between signals and variables

7.1 Styles of Modelling

- **Structural**
  Hierarchical arrangement of interconnected components
- **Behavioural**
  Functional representation of action and timing
  - Data Flow
    behaviour is shown as a series of register assignments
  - Algorithmic
    behaviour is described in terms of programmatic transformations

7.1 Styles of Modelling

Models in VHDL can be constructed in two distinct styles:

1. Structural
   Models in this style are a hierarchical arrangement of interconnected components. They resemble a 'netlist' in that they are a list of interconnected components. Separate models exist to describe the functionality of these components.
2. Behavioural
   A behavioural model is the functional representation of the action and time of the entity.

N.B. A behavioural model must exist for each component used in a structural model.

It is possible to create a hybrid model consisting of both structural and behavioural parts. Such models will have some functionality built into them and will use components to draw functionality from other models.

Behavioural models can be further classified by the structures used in their construction into two groups:

A. Data flow
   The functionality (behaviour) of the entity is shown as a series of register assignments. The models which have been created in the previous chapters of this volume may be classified as data flow models. They show assignment of values to signals.

B. Algorithmic
   The functionality (behaviour) of the entity is described in terms of programmatic transformations. Models constructed in this style are similar to Behavioural Language Models. The characteristic which can be used to identify this style in VHDL is usually the use of **if** statements. Just as hybrid models can be created by mixing structural and behavioural modelling styles, models can be created which have both data flow and algorithmic characteristics.

The reason these concepts have been introduced at this point is in order to lead into the idea of sequential modelling. Algorithmic structures tend to be programmatic - in the software sense. This means that, like programmes, they follow the sequential execution of statements instead of the concurrent execution which has been described so far for VHDL. In this chapter, VHDL structures which permit the use of sequential statements, and sequential statements themselves, are introduced.

7.1.1 Algorithmic nand_gate Model

```
library IEEE ;
use IEEE.Std_logic_1164.all ;

entity nand_gate is
    port (a, b : in Std_logic;
                c : out  Std_logic) ;
end nand_gate ;

architecture algorithm of nand_gate is
begin
    process
    begin
        if  a = '1' and b = '1' then
                c <= '0' after 5 ns ;
    else
                c <= '1' after 10 ns ;
    end if ;
    wait on a, b ;
    end process ;
end algorithm ;
```

7.1.1 Algorithmic nand_gate Model

The model shown in Illumination Box 7.1.1 is another approach towards emulating the functionality of a nand_gate. It should be stressed that there are many ways in which the functionality of an entity can be modelled. Each of these may have some advantage or some disadvantage. The models presented in this volume are written in order to stress particular VHDL constructs. They are not meant, necessarily, to be 'ideal' patterns upon which to base modelling practices.

A glance at the architecture body of this model is enough to see that it contains many new VHDL constructs. Among these are a **process** statement, an **if** statement and a **wait on** statement. Each of these statements is discussed in this chapter, but before formally introducing them let us examine what the model appears to do.

We know that a nand_gate produces a '0' when both inputs to it are '1', and a '1' when either, or both, inputs are '0'. If we read the model like a story this is what it tells us: **begin** a **process**.

Within the process **begin** by checking to see **if** the value in signal *a* is equal to '1' **and** the value in signal *b* is equal to '1'. If they are both '1', assign the value '0' to signal *c* **after** waiting for 5 nanoseconds, **else**, when they are not both '1', assign the value '1' to signal *c* **after** waiting for ten nanoseconds. Finally, **wait** for *a* or *b* to change before carrying on with the process.

The above 'story' adds enhancement to the idea that VHDL is both human and machine readable since the actual VHDL which represents the verbal description given above is essentially a terse version of the 'story'.

Now that, generally, we understand what is going on in the model, the following sections will add depth to the understanding of the specifics of the statements.

7.2 Process Statement

- **Concurrent statement which delineates set of sequentially executed statements**

- **Format**

```
[label :]    process
             -- declarative statements
             begin
             -- sequential activity statements
             end process [label] ;
```

- **Example**

```
la_bel :     process
             begin
                  inter_mediate <= a and b ;
                  c <=  not intermediate ;
                  wait on  a, b ;
             end process la_bel ;
```

7.2 process Statement

The process statement is one of the six concurrent statement types which can be used to construct the architectural body of a VHDL model. It is a concurrent statement which delineates a set of sequentially executed statements.

N.B. Process statements are executed concurrently but statements within a process statement are executed sequentially.

The structure of a process statement is similar to that of the architecture body. It begins with a line identifying the statement (**process**). This is followed by an area where declarative statements can be placed. The declarative area ends with the beginning (**begin**) of the activity or process statement area. This area ends with the end of the process (**end process**).

```
[label :] process
-- declarative_statement area
begin
-- process_statement area
end process [label] ;
```

The square brackets around the word label indicates that the description 'label' is optional. However, as discussed with reference to Illumination Box 7.2.1 the recommendation is that it should always be used.

Statements in the activity, or process, area of the process statement are executed sequentially. Concurrent statements cannot appear within this area. There are a number of sequentially executed statements in VHDL which can be used within this area but it should be kept in mind that none of the six concurrent statement types can appear in the activity area of a process statement.

Look at the example in Illumination Box 7.2. If concurrent statements cannot be used within process statements, what are the signal assignment statements in the example illustrated? The answer is that certain statements, such as signal assignment statements, have both a concurrent and a sequential form. The signal assignment which appears in the example is the sequential form. In this case it is impossible to tell the sequential form from the concurrent form except by the context in which it is used.

7.2.1 Labels

- **Provide internal documentation**
- **User may supply labels for**
  - **Concurrent Assertion Statements**
  - **Concurrent Procedure calls**
  - **Concurrent Signal Assignments**
  - **process Statements**
  - **Loop Statements**
- **user must supply labels for**
  - **Block Statements**
  - **Component Instantiation Statements**
- **System supplies default labels if user does not give labels**
- **Example**

```
la_bel:     process
            begin
                a <= b ;
                wait on  b ;
            end process la_bel ;
```

7.2.1 Labels

The format for a process statement as introduced in Illumination Box 7.2 showed where an optional label could be placed. A label is a distinctive name which identifies a particular unit of functionality. The function of a label is to provide internal documentation. Labels make blocks of functionality easier to locate and, once located, to mark the beginning and end of a block. Labels are very important when debugging a design and their usage is covered throughout the remainder of this volume.

In VHDL, labels must be used on block statements and on Component instantiation statements.

Labels are optional on:

1. Concurrent Assertion Statements
2. Concurrent Procedure Calls
3. Concurrent Signal Assignment Statements
4. Process Statements
5. Loop statements

It is recommended that labels be used on all statements which permit their use. Usually, if the modeller does not provide a label, any CAD system being used will assign a default label for internal tracing purposes but, while the system will be able to use them satisfactorily, the modeller may have difficulty relating to such labels.

Label names follow the same naming rules as objects. The example in Illumination Box 7.2.1 shows the use of a label called "la_bel". This particular example is interesting because it has exactly the same functionality as the following concurrent statement:

```
la_bel : a <= b ;
```

In fact, a concurrent signal assignment statement is interpreted by a CAD system as a process.

Note the use of the label in the concurrent assignment statement. Although this has not been done in the text so far, it is an allowable use of a label.

7.3 Sequential Operations

- **Statements within processes are executed in the order in which they are written**
- **Example**

```
architecture example of  "process" is
begin
        out_1 <= c xor b ;
        pro_exam : process
                variable int : Std_logic ;
                begin
                        int := a and b ;
                        out_2 <= not int ;
                        wait on  a, b ;
                end process pro_exam ;
end example ;
```

7.3 Sequential Operations

"Sequential" means that operations are arranged so that one comes immediately after another. In the previous discussions the activity, or process, area in the process statement was described as a place where sequential statements could be placed. This implies that statements within processes are executed in the order in which they are written.

Concurrent statements are executed in any order. Sequential statements are evaluated in the order in which they are written. Statements after the **begin** in a process statement are evaluated one after the other in the order in which they appear. Once the concurrent process statement is activated, the statements within it are evaluated in sequential order.

Illumination Box 7.3 displays several interesting features:

1. The architecture contains two concurrent statements

   out_1 <= c **xor** b ;
   pro_exam : **process** . . . . . . **end process** pro_exam;

2. A user supplied label (pro_exam) appears on the second concurrent statement (the process statement). The label on the first concurrent statement is not supplied so it defaults to a system default label.
3. We have learned that there is a declarative statement area in the architecture between the first line (**architecture**...) and **begin**. There is also a declarative area in process statements between the first line (**process**...) and **begin**.

   In the example, a variable is declared in the declarative area of the process statement.

4. The body (activity area) of the process statement is made up of sequentially executed statements.

One of the statements in this area in the example is **wait on**. This is a process control statement and is very important. Process statements without some form of process control would execute continuously.

The introduction of process statements opens up numerous questions regarding sequential operations. These questions will be addressed in this and future chapters of this volume.

7.3.1 Sequential Statements

wait Statement
Variable Assignment
signal Assignment*
if Statement
case Statement
Loops
next Statement
exit Statement
return Statement
null Statement
Procedure Call*
assertion Statement*

* execute sequentially or concurrently

7.3.1 Sequential Statements

There are more sequential statements in VHDL than there are concurrent statements. The list in Illumination Box 7.3.1 shows the sequential statements which are discussed in this volume. Some of the statements listed have both a sequential and a concurrent form. Those statements which are indicated with an asterisk may appear in both concurrent statement areas and in sequential statement areas.

The remainder of the statements can appear only in sequentially evaluated areas. At this point, the only sequentially evaluated area of which we are aware is within a process statement. In the next few chapters the use of sequential statements is limited to the activity area within a **process** statement.

Many constructions in VHDL can only appear in specific areas within the model. Appendix C contains a "language construct tree". This tree shows, graphically, where many of the commonly used VHDL constructs may be placed. If in doubt as to whether a certain statement may, or may not, be used in a particular place in a model, refer to the language construct tree.

7.4 wait Statement

- **Provides execution control of process statements**

- **Without control, statements within process execute continuously**

- **Format**

  wait [on list **][until** condition**][for** time**] ;**

- **Example**

```
process
    variable inter_mediate : Std_logic;
begin
    inter_mediate : a and b ;
    c <= not inter_mediate ;
    wait on  a, b ;
end process ;
```

7.4 wait Statement

Without some type of control a **process** statement executes continually. The sequential statements within such a process would be executed repeatedly. Hence, without some sort of control, a **process** statement creates an unending loop. For this reason, **process** statements cannot be written without control statements. The **wait** statement is one type of process control.

In the discussion on concurrent signal assignment statements we learned that the statement "activates" when a signal on the left of the assignment operator changed. The statement was said to be "sensitive" to the signals on the left of the operator.

Process statements are continually active. Therefore, instead of a mechanism for activating the statement, there must be some method for providing execution control (i.e. inhibiting continual activity). The **wait** statement provides such a mechanism for controlling the activity of process statements.

The format of the wait statement provides three options. At least one of the options must be chosen:

wait [**on** list] [**until** condition] [**for** time] ;

As might be expected from the first word in the statement, the purpose of the wait statement is to cause the execution to wait upon some factor. Such factors include signal changes, specific conditions and time. They are discussed in more detail in Illumination Box 7.4.1.

There is no limit to the number of wait statements that can appear in a process statement, but there should be at least *one* wait statement. Without some kind of process control a process statement would execute continually.

A **process** statement without process control should produce an error signal in a compilation package.

7.4.1 wait Control

- **Suspension may be based on**
  - **signal sensitivity**
  - **condition**
  - **timeout**

- **More than one suspension criterion may be included in wait**

- **Format**

  wait [on list**][until** condition**]for** time**] ;**

- **Examples**

  wait ;
  wait on a ;
  wait on b **for** 100 ns ;
  wait until a = '1' **for** 50 ns ;
  wait on a **until** (c = '1') **for** 50 ns ;

7.4.1 wait Control

In the previous section we learned that the **wait** statement can be used to control processes. A **wait** without options suspends evaluation of the process for ever. Therefore, the wait statement is usually written with one, or more, of the following modifiers included in the statement:

1. Signal sensitivity

   The evaluation of a process may be suspended until there is a change in the waveform of a signal. This is somewhat analogous to what takes place in a concurrent signal assignment statement. When a signal on the left side of the operator changes, the statement is activated and evaluated. A **wait** statement with the **on** option in a process statement suspends evaluation until there is an event signal in the **on** list.

   The following statement would suspend evaluation until signal *a* changed value:

   wait on a ;

2. Condition

   A condition invokes one of the types of relational comparison involving a signal or signals. The evaluation of the process is suspended until the waveform of the signal(s) in the condition satisfies the condition being checked. Any of the relational operators introduced in Chapter 5 (Section 5.4) may be used in the condition (=, /=, <, <=,>,>=). When the condition evaluates as **true** the **wait** statement allows evaluation to continue. If only the **until** option appears in the **wait** statement the condition is checked whenever there is an event on any signal in the condition.

   The following statement would suspend evaluation until the waveform of signal *a* is '1':

   wait until a = '1' ;

   If *a* equals '1' *when* this statement is encountered the process is suspended until *a* changes to some other value and *then* changes back to '1'.

3. Timeout

   A third way to control the process is by suspending evaluation for a specific period of time. The **for** option in the wait statement identifies that evaluation is to be suspended to a given time period.

   The following statement would suspend evaluation for 100 ns:

   wait for 100 ns ;

7.4.2 Gated d_latch

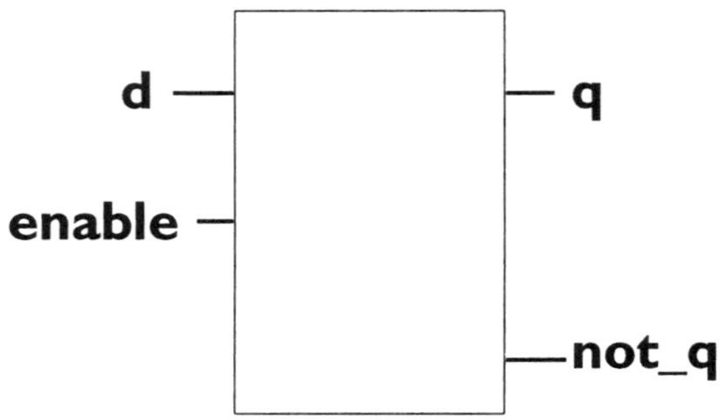


```
architecture example of d_latch is
begin
  latch : process
          begin
            wait on d, enable until enable = '1'
            q <= d ;
            not_q <=  not d ;
          end process latch ;
end example ;
```

It is possible to use more than one of the conditions on a wait statement. When **on** is used with **until,** evaluation is suspended until the signal in the **on** changes and the condition is true. When **on** is used with **for**, evaluation is suspended until the signal changes *or* the time limit is reached. When **until** and **for** are used together, evaluation is suspended until the condition is met *or* the time limit is reached. When all three conditions are used, evaluation is suspended until the signal changes *and* the condition is met *or* the time limit is reached.

7.4.2 Gated d_latch

The model shown in Illumination Box 7.4.2 shows how a process statement could be used to construct a simple functional representation of a gated d_latch. For this discussion the following specification of a gate d_latch is used.

Input to the device consists of two signals: an "enable" signal and a data line signal "d". Output is two signals: a signal called "q" which assumes the value of the data line when the "enable" signal is '1' and a signal called "not_q" which assumes the inverse of the data line when the "enable" signal is '1'. The data line signal and the "enable" signal are both of type BIT ('0' or '1').

The architecture contains a single process statement. Within the process statement are three sequential statements; a wait statement for process control and two signal assignment statements.

Look at the wait statement:

wait on d, enable **until** enable = '1' ;

The sensitivity list following **on** contains two signal names, *d* and *enable*. Whenever either of these signals changes value the wait statement will check to see if the value of the *enable* waveform is equal to '1'. If it is, execution will be allowed to continue and the two signal assignment statements will be evaluated.

Execution of the process statement does not end when the '**end process** latch;' statement is reached. Instead, control returns to the beginning of the process and evaluation is done again. In this case the first statement encountered when the second evaluation of the process is begun is the wait statement. This statement suspends evaluation until either of the two signals in the **on** list change and the enable signal is equal to '1'.

7.5 process Sensitivity List

- **Equivalent to a wait on statement**
- **If used, wait statements may not be used in process**
- **Format**

```
[label :] process (sensitivity list)
        begin
            -- sequential statements
    end process [label] ;
```

- **Example (Equivalent)**

```
process (a, b)          {  process
begin                   {  begin
c <= a and b ;          {  c <= a and b ;
end process             {      wait on a, b ;
                        {  end process        }
```

Notice that the wait statement in this example appears immediately after the **begin** in the process statement. In other examples the wait statement was placed immediately before the **end process**. Wait statements may be placed anywhere within the activity area of the process. However, the placement of the wait statement may have a marked effect when a simulator or synthesis tool is invoked on the model.

As an example of the above, when Quicksim® is invoked on a model all the statements in the model are subject to evaluation. Normally, all concurrent statements are evaluated. However, whether the sequential statements within a process are evaluated, or not, depends upon the placement of the wait statement. If the wait statement appears immediately after the **begin** the sequential statements will not be evaluated during invocation. When execution of the process reaches the wait statement it is suspended. Thus, statements following the wait will not be evaluated. If the wait statement appears immediately before the **end process** all the sequential statements in the process will have been evaluated before the execution is suspended.

Notice also that both signal *d* and *enable* appear in the **on** list. If only signal *d* appeared in the list the model would not perform correctly. With only *d* in the list the output would only change when signal *d* changed and the value of the enable was '1'. No change would occur when the signal *enable* changed to a '1' and signal *d* did not change.

7.5 process Sensitivity List

A **process** sensitivity list may sometimes be used in place of a wait statement to control process execution. This list is equivalent to a **wait on** statement. The sensitivity list contains the signal names which would have appeared in the list following **on**. The list appears in parentheses immediately following the key word **process**. For example:

```
example_1 : process (a, b)
    begin
            c <= a and b ;
    end process example_1 ;
```

In this example the "(a, b)" is the sensitivity list. It is equivalent in function to a **wait on** statement appearing just before the **end process**; i.e.:

```
example_2 :process
    begin
            c <= a and b ;
            wait on a, b ;
    end process example_2 ;
```

7.6 variable Definition

- **Used only within sequential areas**
- **Format**

  variable variable_name : **type** [:= initial_value] ;
- **Variables assume value instantly (no time dimension)**
- **Assignment operator**

  :=
- **Example**

```
example :  process
                variable tla : Std_logic:= '0' ;
           begin
                tla := a and b ;
                out_2 <= tla xor c ;
                wait on a, b ;
           end process example ;
```

A major drawback to the sensitivity list is that, once it is used, **wait** statements cannot be used in the process. For example, the following is illegal:

```
example_3 :process (a, b)
    begin
        c <= a and b ;
        wait for 10 ns ;   -- this is illegal
    end process example_3 ;
```

This example shows how an inexperienced modeller might attempt to add "delay" to a process sensitive to signals *a* and *b*. This an illegal usage of a sensitivity list and a **wait** statement in a single process statement.

As a footnote it is interesting to observe that the concurrent signal assignment statement:

```
c <= a and b ;
```

is equivalent to the sensitivity list example with the label "example_1".

7.6 variable Definition

In Chapter 5 (Section 5.1) we were introduced to three classes of objects; constants, variables and signals. Variables are objects which can be assigned a value which may be changed during the course of the simulation of the model. A variable is a container in which a single value can be stored. There are several differences between variables and signals:

1. Variables may be declared and used only within sequential areas of a model
   Signals can be declared only within concurrent areas of the model but they may be used in both concurrent and sequential areas.
2. Variables assume a value instantly.
   Signals assume a value after some period of delay. If no explicit delay is stated the minimum delay is one delta time period, which emulates concurrent execution.
3. Variables cannot be used outside the sequential functional block in which they are declared. There is no such thing as a global variable. Signals are "global" in the sense that they can be used throughout their scope.
4. The variable assignment operator is the symbol :=. Although this symbol looks like the initial_value assignment operator it is not the same. The signal assignment operator is the symbol <=.
5. Signals carry both a value and a time at which the value takes effect. Variables carry only a value since they take effect immediately. There are no "drivers" for variables.

7.6.1 signal vs variable Assignment

```
var_example : process
        variable num, sum : integer := 0 ;
    begin
        wait for 10 ns ;
        num := num + 1 ;
        sum := sum + num ;
    end process var_example ;

architecture x of y is
        signal num, sum : integer : 0 ;
begin
sig_example : process
    begin
        wait for 10 ns ;
        num := num + 1 ;
        sum := sum + num ;
    end process sig_example ;
end x ;
```

Since variables can be declared only within sequential areas, and since the only sequential area mentioned thus far is in the process statement, variables may be defined within process statements.
The declaration format is:

variable variable_name : **type** [:= initial_value] ;

Except for the word "**variable**" at the beginning variables are declared in the same manner as internal signals. The "type" can be any of the predefined types which have been discussed. The initial value is optional. If it is not given the default is the left most value in the type's value set.

7.6.1 signal vs variable Assignment

The assignment of values to signal objects and variable objects can produce very different results in otherwise identical situations.

Consider the two process statements in Illumination Box 7.6.1. The "action" is the same in both processes. "1" is added to an object and the sum of the values of the object is accumulated over time. A **wait** statement provides execution control by suspending execution for 10 ns each time through the process. The difference in the two processes is that, in one process, variable objects are used to hold the values and, in the other, signal objects are used to hold the values.

In each of the processes the objects are of type integer and have initial values of zero. Since signals cannot be declared within processes we must assume that the values of *num* and *sum* in sig_example are declared in the declarative region of the architecture.

The following tables represent the values of *num* and *sum* in each of the processes during a period of 30 ns:

var_example

time	num	sum
0	0	0
10	1	1
10+	1	1
20	2	3
20+	2	3
30	3	6
30+	3	6

sig_example

time	num	sum
0	0	0
10	0	0
10+	1	0
20	1	0
20+	2	1
30	2	1
30+	3	3

7.7 Sequential Signal Assignment

- **Without transport phrase only last value assigned to signal within sequential assignment is recognised**

- **Example**

```
no_transport : process
    begin
        wait on c ;
        out_1 <= '1' after 10 ns ;
        out_1 <= '0' after 20 ns ;
    end process ;

with_transport : process
    begin
        wait on c ;
        out_1 <= '1' after 10 ns ;
        out_1 <= transport '0' after 20 ns ;
    end process;
```

The differences in the results are due to the time at which values actually take effect. Variable values are assigned at the time the statement is evaluated. Signal values are assigned at least one delta time period in the future. Therefore, each time the variable *num* changed in var_example its value was assigned immediately and was available to be added to the variable *sum*. In sig_example the value of *num* changed but was not assigned until one delta time period later. Therefore, the value added to *sum* was the value of *num* at the beginning of the delta time period, rather than the changed value.

One delta time, or iteration, passes for a process each time its execution is suspended (the **wait** statement is encountered).

7.7 Sequential Signal Assignment

In a previous chapter we learned how delay affects concurrent signal assignments; specifically, how the filtering effect of the **after** clause ignores signals with durations less than the amount of delay and how this can be overcome with the **transport** option. We also learned that resolution functions are required for signals which receive values from multiple concurrent signal assignment statements (multiple drivers).

There are similar rules which apply to sequential signal assignment statements:

1. Within a single process a signal may have more than one sequential signal assignment statement assigning values to it without there being a resolution function.
2. Only the last value assigned to a signal with sequential signal assignment statements is recognised.
3. When the **transport** option is used sequential signal assignment statements assignments which are scheduled for times before the time in the statement are retained. Assignments which are scheduled after the time in the statement are deleted.

The example given in Illumination Box 7.7 demonstrates these three points. The trace for the no_transport example would be:

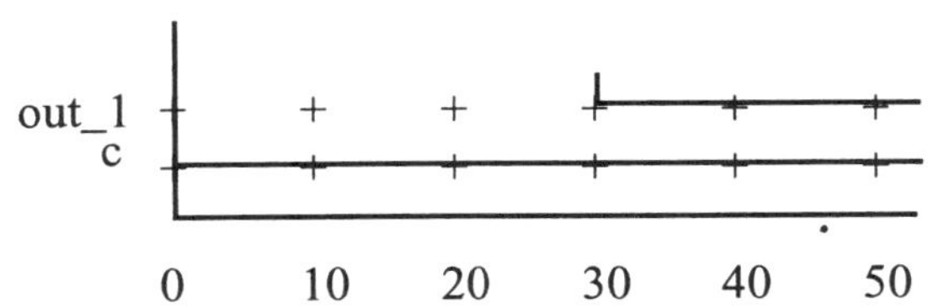

Figure 7.1(a) No-transport option

When the transport option is used (as in the second part of the example), the value of out_1 assigned to occur at 10 nanoseconds would not be lost.

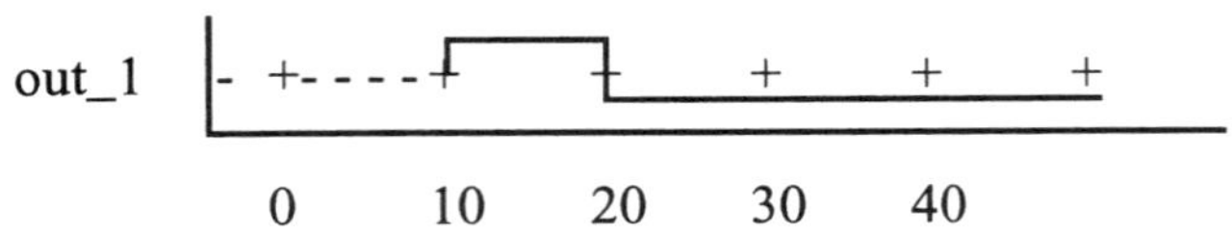

Figure 7.1(b) With_transport option

7.8 Exercises

7.8.1 Describe the effect of the **wait** statement in the following description

```
architecture waiting of sig_name is
signal y : Std_logic := 'Z' ;
begin
    process
      begin
          y <= '1' after 10 ns ;
          wait ;
    end process ;
end waiting ;
```

What would be the effect of adding the line

```
y <= transport '0' after 5 ns ;
```

after the existing statement for y in the description?

7.8.2 Explain the difference between the use of the **process** sensitivity list and the **wait on** statement.

7.8.3 Does it matter where the **wait on** statement is placed in the process, e.g. before the **end process** or after the **begin?**

7.8.4 i) When does a variable assignment take effect?
ii) When are signals which are assigned new values actually updated?

7.8.5 Create a model which accepts two inputs and produces one output. If both inputs are high, the output is to be low. If either input is low, the output is to be high. If an input is 'X or 'Z' and the other input is high, or unknown, the output is to be 'X'. All output is to be delayed by 25 ns.

7.8.6 Show how signal values shorter than, say, 10 ns can be removed from a waveform. If such signals are to be retained in a waveform which is to be delayed by 10 ns what change must be made to the description?

7.8.7 Create a model for the evaluation of the first 16 bits of the Fibonacci sequence.

Chapter 8

Sequential Modelling and Attributes

Overview

This chapter is a "building" chapter. Two major new VHDL concepts, and the application of these concepts to modelling situations, are introduced and discussed. These are sequential modelling and how to control sequential code and signal attributes and their use in modelling such processes as setting up edge triggering, setup time checks and hold time checks.

In this chapter you will learn

- To write sequential code
- How to use **if** statements
- How to use attributes

8.1 Sequential Modelling

```
library IEEE ;
use IEE.Std_logic_1164.all;

entity nand_gate is
    port (a,b : in BIT;
              c : out BIT);
end nand_gate ;

architecture algorithm of nand_gate is
begin
    process (a, b)
    begin
        if  a = '1' and b = '1' then
            c <= '0' after 10 ns ;
        else
            c <= '1' after 10 ns ;
        end if ;
    end process ;
end algorithm ;
```

8.1 Sequential Modelling

The model shown in Illumination Box 8.1 is yet another approach to representing the functionality of a nand_gate in VHDL. The style used in this model is sequential modelling.

The entity declaration in this model is very similar to the entity declaration in previous models of the nand_gate. The only difference is that the ports have been declared to be type BIT rather than type Std_logic. The reason for this is to decrease the complexity associated with modelling multiple states. For this model the only signal values that are possible are '1' and '0' (therefore, is it necessary to have LIBRARY and USE statements in this model? The reader may decide).

Before discussing the architecture let us assume that the following specification represents the functionality which will be modelled in this example;

Create a model of a device which has two inputs and one output. The output value changes 10 ns after a change in the input values. If both inputs are '1' the output is to be '0'. Any other combination of '1' and '0' on the input is to produce an output of '1'.

Now let us consider the architecture of the model:

1. The body of the architecture consists of one process statement.
   A functional architecture consists of one, or more, concurrent statements. A process statement is a concurrent statement and so this requirement is met.
2. The process statement is "sensitive" to signals *a* and *b*.
   Signals *a* and *b* appear in the sensitivity list associated with the process statement. Therefore, whenever there is an event (a change in value) on either of these signals the process statement will be executed.
3. When the process is executed the values of signals *a* and *b* will be checked.
   If the value of both signals is '1' a value of '0' is assigned to the output signal *c* after a delay of 10 ns. If any other combination of '1' and '0' occurs on the input signals a value of '1' is assigned to the output signal.
4. Once the signal assignment is made the execution of the process statement is suspended until either, or both, of the input signals changes value.

The sequential model thus described reinforces the idea of human and machine readable code in that it can almost be "read" in the same manner as the model specification, i.e. the **if** statement follows the "spoken" logic of "if" in English.

This model is simpler than when the type of **port** is Std_logic. For example, it does not account for 'X' and 'Z' states. If they were to occur they would cause the same output as any state except '1' and '1'. This is obviously not acceptable;

8.2 Controlling Sequential Code

- **Conditional control**
  - **if Statements**
  - **case Statements**
- **Iterative control**
  - **Loops**

however, to show how the **if** statement can be used to handle more complex types it is necessary to understand more about the **if** statement.

8.2 Controlling Sequential Code

When process statements were introduced it was pointed out that they required some type of execution control. Without this control a process statement executes continually. The **wait** statement was introduced as a method of providing execution control. It provides a method of determining when the process takes place or when segments of the process take place (if multiple **waits** are used within the same process statement). It does not provide a good method of determining which statements are executed or how many times they are executed.

The control of sequential code requires the ability to select which statements within the sequential series to execute and which not to execute and to determine how many times a statement, or series of statements, in the code is executed. Control of statement execution is referred to as "conditional control". The number of times a statement is executed is referred to as "iterative control".

The repetition of a set of statements is called a loop. There are three types of loop in VHDL:

- **for** loops
- **while** loops
- **loop - exit** constructs

Loops are discussed in Chapter 10 of this volume.

Conditional control requires the ability to determine whether a statement is, or is not, executed. This requires some method of making choices based on existing conditions. In VHDL sequential code there are two general methods for doing this:

- **if** statements
- **case** statements

if statements are introduced and discussed in this chapter. **case** statements are introduced and discussed in Chapter 9.

8.2.1 if Statement

- **Provides conditional control of sequential statements**
- **The condition in the statement must evaluate to a Boolean value**
- **Execution of statements within if occurs when condition is TRUE**
- **Terminates with end if**
- **Format**

```
if condition then
    --sequential statements
end if ;
```

- **Example**

```
if a = '1' then
    count := count + 1 ;
end if ;
```

8.2.1 if Statement

The **if** statement provides conditional control of sequential statements. It is itself a sequential statement and can only appear in sequential statement areas. This statement has several variations but the basic form always follows the format:

```
if condition then
-- sequential_statements
end if ;
```

The 'condition' in the first line of the **if** statement must evaluate to a BOOLEAN value (TRUE or FALSE), therefore, this can be an object of type BOOLEAN or it can be a relational comparison between two objects of the same type. The objects can be signals, variables or constants as long as they are of the same type (or the relational operators have been overloaded to handle objects of differing types).

The sequential statements contained within the **if** statement are evaluated if, and only if, the condition is TRUE. If the condition is not true execution skips to the end of the **if** statement and resumes evaluation with the first statement following the end.

The end of the **if** statement is marked by the phrase **end if** ;. Note that this is written as two distinct words. This statement should not be confused with the ENDIF of other languages since writing the statement in this way will cause an error during compilation.

The example shown in Illumination Box 8.2.1 illustrates the use of an **if** statement. It is evaluated as follows:

- The value stored in object *a* is compared with the character literal '1'. The result will be either TRUE OR false.
- If the result of the comparison is TRUE, the statement count := count +1 ; is evaluated (one is added to the value stored in the variable *count*).
- If the result of the comparison is FALSE, the statement count := count + 1 ; is skipped (i.e. not evaluated).

In either case the series of statements under the control of the **if** statement is terminated by the **end if** ; and normal evaluation is resumed with the next statement after the end of the **if** statement.

8.3 Relational Operators

- **Return Boolean values as their result**

Operator	Operation
=	equal
/=	not equal
<	less than
<=	less than or equal
>	greater than
>=	greater than or equal

8.3 Relational Operators

Relational operators perform comparisons to determine the relationship between two items. The two items can be two objects of the same type or an object and a value set of the object's type. The result of the comparison will always be a BOOLEAN value (TRUE or FALSE).

The VHDL relational operators are:

1. =
   Equal. TRUE is returned when the values on the left and right sides are identical.
2. /=
   Not equal. TRUE is returned when the values on the left and right side are different.
3. <
   Less than. TRUE is returned when the value on the left side is smaller than the value on the right side. In numeric and physical comparisons "smaller" means that the value occurs earlier in the type's value set list, For example, in Std_logic ('X', '0', '1', 'Z'), 'X' is the "smallest" and 'Z' the "largest".
4. <=
   Less than or equal. TRUE is returned when the value on the left side is smaller than or equal to the value on the right side.
5. >
   Greater than. TRUE is returned when the value on the left side is larger than the value on the right side. Using Std_logic as an example, '0' is greater than 'X'.
6. >=
   Greater than or equal. TRUE is returned when the value on the left side is equal to or larger then the value on the right.

If objects are compared with relational operators the type of the objects must be the same. However, the class of objects can be different. For example, a signal can be compared with a variable or a constant if the types are the same. Relational Operators are valid for all types. They can be used with any of the predefined types or with any user defined types. They are, in a sense, overloaded for all types.

8.3.1 if - else Construction

- **Provides selection control between two groups of statements based on a single condition**
- **Format**

```
if condition then
        -- sequential statements
else
        -- sequential statements
end if ;
```

- **Example**

```
if a = '1' then
        ones_count := ones_count + 1 ;
else
        others_count := others_count + 1 ;
end if ;
```

8.3.1 if-else Construction

One of the variations of the **if** statement is the **if-else** construction. This form of the **if** statement provides selection control between two groups of statements based on a single condition. This is a useful construction because it permits an alternative to the TRUE evaluation of the condition. Consider the following statement:

```
if a = '1' then
c := c + 1 ;
end if ;
b := b - 1 ;
```

The statement c := c + 1 ; is executed only when the value in *a* is equal to '1'; but the statement b := b - 1 ; is executed whether or not *a* is equal to '1'. Suppose that the modeller wanted one added to *c* only if *a* was equal to '1' and wanted one subtracted from *b* only if *a* was not equal to '1'. One way to do this would be:

```
if a = '1' then
c := c + 1 ;
end if ;
if a /= '1' then
b := b - 1 ;
end if ;
```

While this does achieve the desired result it is somewhat clumsy. A better way would be:

```
if a = '1' then
c := c + 1 ;        -- this is done if condition is TRUE
else
b := b - 1 ;        -- this is done if condition is FALSE
end if ;
```

This represents the **if-else** construction. When the condition evaluates to TRUE the sequential statements *before* the **else** are executed and the statements after the **else** are ignored (skipped over). When the condition evaluates to FALSE the sequential statements before the **else** are ignored (skipped over) and the statements *after* the **else** are executed. The effect is to do either "this" or "that" based on the result of the condition.

8.3.2 elsif Construction

- **Provides multiple alternatives based upon multiple conditions**
- **Format**

```
if condition then
        -- sequential statements
elsif condition then
        -- sequential statements
end if ;
```

- **Example**

```
if a = '1' then
        ones_count := ones_count + 1 ;
elsif a = '0' then
        zeros_count := zeros_count + 1 ;
else
        others_count := others_count + 1 ;
end if ;
```

8.3.2 elsif Construction

Another useful variation of the **if** statement is the **elsif** construction. This form of the **if** statement provides selection control between groups of statements based on multiple conditions. The major difference between this variation and the **if-else** variation is that in the **if-else** all factors which cause the condition to evaluate as FALSE produce the result (execution of the statements following the **else**) and in the **elsif** each of the factors can be tested for in succession.

For example, suppose that if the value of a Std_logic signal were 'X' a certain series of statements were to be executed. However, if the value were '0' another series of statements were to be executed; if the value were '1' yet another series of statements were to be executed and if the value were 'Z' a fourth and final set of statements were to be executed. This could be mapped as follows:

```
if a = 'X' then
b := b + 1 ;
elsif a = '0' then
c := c + 1 ;
elsif a = '1' then
d := d + 1 ;
end if ;
```

Actually the final **elsif** could be replaced by an **else** since all possible alternative values of a Std_logic object have been accounted for except 'Z'. This introduces several interesting points.

- An **else** can only be used as the last item in an **elsif** list. Since **else** is all inclusive there can be no alternative checks after it.
- **else** is always associated with the last **elsif** condition.
- The condition in an **elsif** does not have to check the same object that was checked in previous conditions.
- Only the statements under the first **elsif** that evaluates as TRUE are executed. Thus, even if other conditions would evaluate as TRUE, their underlying statements are not executed.
- **elsif** does not produce the same results as "nested" **if** statements. For example, these statements do not produce the same result:

```
if b = '0' then              if b = '0' then
        b := b + 1 ;                 b := b + 1 ;
elsif a = '1' then               if a = '1' then
        c := c + 1 ;                 c := c + 1 ;
end if ;                         end if ;
                             end if ;
```

8.4 Algorithmic nor_gate

```
library IEEE ;
use IEEE.Std_logic_1164.all ;

entity nor_gate is
      port (a, b : in Std_logic ;
                  c : out Std_logic ) ;
end nor_gate ;

architecture algorithm of nor_gate is
begin
      nor_process :
      process (a, b)
      begin
            if  a = '0' and b = '0' then
                  c <= '1' after 25 ns ;
            elsif (a = '1') or (b = '1') then
                  c <= '0' after 25 ns ;
            else
                  c <= 'X' after 10 ns ;
            end if ;
      end process nor_process ;
end algorithm ;
```

8.4 Algorithmic nor_gate

The model shown in Illumination Box 8.4 represents another algorithmic approach to a simple gate action. In this case the model attempts to emulate the functionality of a simple nor_gate.

Before discussing the architecture let us assume that the following specification represents the functionality which will be modelled in this example:

> *Create a model of a device which has two inputs and one output. The output value changes 25 ns after a change in the input values if the change is to a '1' or to a '0'. It changes in 10 ns if the change is to an 'X'. If both inputs are '0' the output is to be a '1'. A combination of '1' and any other value produces an output of '0'. All other combinations produce an output of 'X'.*

Now consider the architecture of the model:

1. The body consists of one **process** statement that is sensitive to the two input signals *a* and *b*.
2. When the process is executed the values of signals *a* and *b* are checked.
   · If the value of both signals is '0' a value of '1' is assigned to the output signal *c* after a delay of 25 ns.
   · If both signals do not have a value of '0' each signal is checked to see if it has a value of '1'. Should either signal have a value of '1' a value of '0' is assigned to the output signal *c* after a delay of 25 ns.
   · If none of the previous conditions is satisfied (evaluates as TRUE) a value of 'X' is assigned to the output signal *c* after a delay of 10 ns.
3. Once the evaluation has reached the **end process** nor_process execution of the process statement is suspended until either or both of the input signals *a* and *b* change again.

As with the sequential nand_gate model this model is a "human readable" statement as well as an machine executable statement.

It should be noted, however, that models depicted in this text are designed to show certain constructions and are not necessarily the most efficient or "shortest" ways to construct a model.

Notice the use of relational and logical operators in this model. When these operators are mixed in a statement the relational operators are evaluated first and then the logical operators are evaluated. Note also that parentheses have been placed around the relational comparisons in the **elsif**.

8.5 Modelling Enables

- **Enables activate devices or put the devices in operating mode**
- **Example**

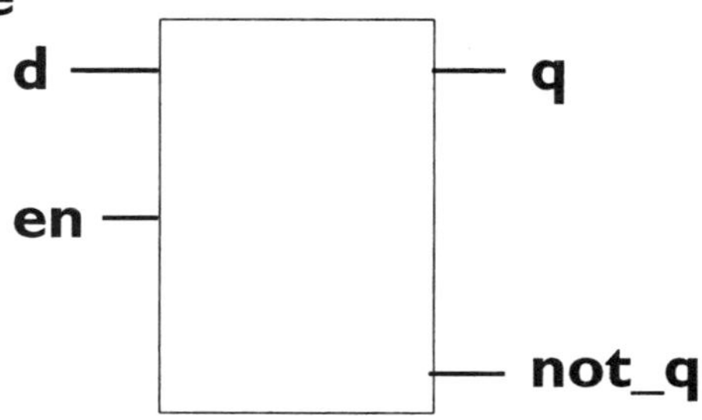


```
architecture algorithm of gate_d_latch is
begin
        d_latch :
        process (en, d)
        begin
            if en = '1' then
              q <= d after 25 ns ;
              not_q <= not  d after 25 ns ;
            end if ;
        end process d_latch ;
end algorithm ;
```

This was done for "readability" and not of necessity. It is interesting to note that **and** and **or** do not have to be overloaded in order that Std_logic objects for this model may be executed. **and** and **or**, in this case, are comparing BOOLEAN values.

8.5 Modelling Enables

An enable is a signal which activates a device or puts the device into an operating mode. The use of process statements makes modelling devices with enable circuitry reasonably straightforward.

Consider the following specification:

> *Create a model of an "enabled" d_latch. There will be two inputs, the enable signal and the data signal: and there will be two outputs, the data signal and the inverse of the data signal. The outputs assume their values from the inputted data signal only when the enable signal is '1'. In other words, the output follows the input when the enable signal is high. Assume a delay of 10 ns in the change of output from the change of input.*

Now consider the architecture of the model in Illumination Box 8.5:

1. The body of the architecture consists of one process statement which is sensitive to both the data signal *d* and the enable signal *en*.
2. When either of the input signals changes, the process is executed
3. If the value of the enable signal is '1', the value of the input signal *d* is transferred to the output signal *q*, and the inverse value of *d* is transferred to the output signal *not_q*.
4. When the evaluation has reached the **end process** d_latch, execution of the process statement is suspended until either, or both, of the input signals *d* and *en* change again.

The structure of this model makes it possible for the value of signal *d* to change, but the values of signals *q* and *not_q* remain the same. The **if** statement controls execution of the signal assignment statements, and the value of the input data signal is not transferred to the output signals unless the enable signal is high. This is exactly what the specification calls for when it states: *output follow input when the enable signal is high.*

8.6 Modelling Oscillators

- **Simple oscillators switch back and forth between two states**
- **Example**

```
architecture algorithm of oscillator is
        signal clk : Std_logic := '0' ;
begin
clock : process
                variable switch : bit := '0' ;
         begin
                if switch = '0' then
                        clk <= '0' ;
                else
                        clk <= '1' ;
                end if ;
                switch := not switch ;
                wait for 25 ns ;
         end process clock ;
end algorithm ;
```

8.6 Modelling Oscillators

An oscillator switches back and forth between two or more states. The most common type of oscillator is a clock. Clocks are usually designed to switch periodically between '1' and '0'. The example in Illumination Box 8.6 shows one approach to modelling a simple oscillator.

Consider the following specification:

Create a model of a simple oscillator. This device is to switch between '0' and '1' every 10 ns.

Now consider the architecture of the model in Illumination Box 8.6:

1. The body of the architecture consists of one process statement that is controlled by a **wait** statement.
2. An internal signal called *clk* is declared in the architecture declarative area and a variable called *switch* is declared in the process declarative area. Both of these objects are set to an initial value of '0'.
3. When, for example, Quicksim is invoked on the model, the process is executed to the **wait** statement. Since the initial value of *switch* is '0' the **if-else** executes the statements before the **else** condition. This causes *clk* to be assigned the value '0'. Execution then skips to the **end if** statement and the statement following is executed. This statement assigns the value '1' to *switch*. When the following **wait** statement is executed, execution is suspend. The reader is left to consider what would happen if an initial value were not assigned to the variable *switch*.
4. When the model is simulated the **process** statement is executed every 25 ns (the **wait** statement). Each time through the process the value of *clk* is set either to '1' or to '0' depending upon the value of *switch*. Since *switch* is inverted (**not**) each time through the process the clock also changes.

Although this is not the most efficient way to model a clock it does demonstrate some interesting concepts.

- *clk* is an internal signal. It does not enter, or leave, the model.
- **not** is used to change *switch* from '0' to '1' and back. This is possible because *switch* is type BIT.
- This process repeats itself every 25 ns for as long as the model is executed.

Clock models would not normally be created in this manner as it is unnecessarily complex.

8.7 Modelling Clocks

- ```
  clock_1 : process
            begin
                clk <= not clk ;
                wait for 25 ns ;
            end process clock_1 ;
  ```

- ```
  clock_2 : process
  begin
      clk <= not clk after 25 ns ;
      wait on  clk ;
  end process clock_2 ;
  ```

- ```
  clock_3 : process(clk)
  begin
      clk <= not clk after 25 ns ;
  end process clock_3 ;
  ```

- ```
  clk <= not clk after 25 ns ;
  ```

8.7 Modelling Clocks

The clock oscillator in Illumination Box 8.6 can be modelled in more efficient ways. The four examples in Illumination Box 8.7 show other ways in which this functionality could be modelled.

The signal *clk* must be declared at some point in the model. It could be declared in the port clause of the entity declaration or it could be declared in the declarative area of the architecture. If *clk* appeared in the port clause the processes shown on the facing page could be in error.

Signals which appear in the port clause are assigned a mode. In past examples this mode has been either IN or OUT depending upon the direction of the data flow relative to the model. Although it has not been stated before it is important to note that *values cannot be assigned to signals of mode IN and cannot be drawn from signals of mode OUT.* This is illustrated in the following model fragments:

```
port (a, b, c, : in INTEGER ;
         d, e, : out INTEGER) ;

c <= a + b ;   -- illegal, c is mode in and therefore cannot be
                  assigned to
d <= e + a ;   -- illegal, e is mode out and therefore cannot
                  be drawn from
```

This example shows that signals of mode **in** cannot be used to the left of the signal assignment operator and that signals of mode **out** cannot be used to the right of the signal assignment operator.

Since the signal *clk* in the examples in Illumination Box 8.7 appears on both sides of the signal assignment operator it cannot appear as either mode **in** or **out** in the port clause. If *clk* must appear in the **port** statement it should be declared as mode **inout**. Signals that are declared within the architecture are automatically of mode **inout**.

The **process** statements in Illumination Box 8.7 produce similar results (the differences are limited to whether the clocking signal starts at '1' or '0'). The signal *clk* oscillates between '0' and '1' every 25 ns. Since the operator **not** is defined for signals of type BIT and since it produces the inverse of the current value ('1' if '0' and '0' if '1') this result would be expected. However, the reader might like to consider whether the same result could be expected if the clock signal were of type Std_logic?

The answer lies in the order in which the values for the overloaded operator **not** is presented in the package Std_logic_1164.

8.8 Modelling Synchronous Devices

- **Changes in output occur on a triggering output**
- **Example**

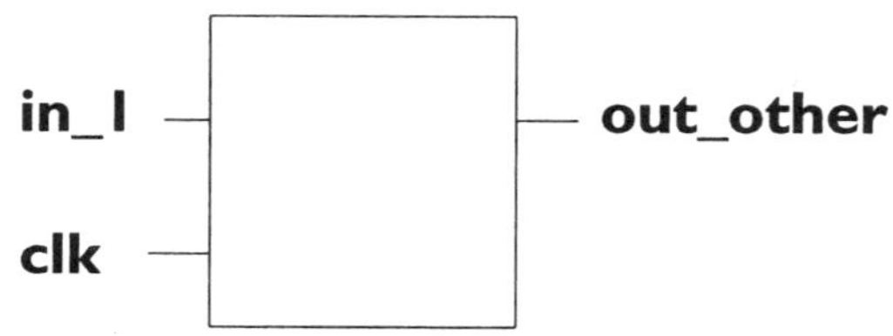


```
architecture example of sync_device is
begin
    in_out :  process (clk)
              begin
                   out_other <= in_1 ;
              end process in_out ;
end example ;
```

8.8 Modelling Synchronous Devices

Synchronous devices have a fixed time relationship. Typically, changes in output occur on an oscillating, triggering input. The difference between synchronous devices and enabled devices is that the triggering input tends to be periodic in the first and sporadic in the second.

The model in the example in Illumination Box 8.8 represents a very simple synchronous device which corresponds to the following specification:

Create a model of a device which transfers the value of the data input line to the output line each time a clocking signal changes.

In this specification the clocking signal is stated to come from outside the device. Hence there are two input signals (the clocking signal and the data line) and one output signal. The architecture in Illumination Box 8.8 satisfies the requirements of this specification.

1. The body of the architecture consists of one **process** statement that is sensitive to the *clk* signal. Any change in the value of this signal causes the process to be executed.
2. When the process is executed the value in the signal *in_1* is assigned to the signal *out_other*.

If the clocking signal is periodic the process causes the change in output to occur at fixed times. Since the changes in output occur on an oscillating, triggering input the device is said to be synchronous.

The clocking signal could originate within the model. The following example shows how two blocks of functionality could be placed in the same model:

```
architecture example of two_process
        signal clk : BIT := '0' ;
begin
        clock : process (clk)
                clk <= not clk  after 10 ns ;
        end process clock ;
        in_out : process (clk)
                out_other <= in_1 ;
        end process in_out ;
end example ;
```

The clock in this example changes value every 10 ns. Each time it changes value the two processes are activated. The clock is set to change again in ten nanoseconds and the value of *in_1* is assigned to *out_other*. Hence the clocking source may appear as a separate model or as a functional block within the model.

8.9 Edge Triggering

- **Changes occur only at a specific point on a triggering input**
- **Typically occurs on rising or falling edge of clock pulse**
- **Example**

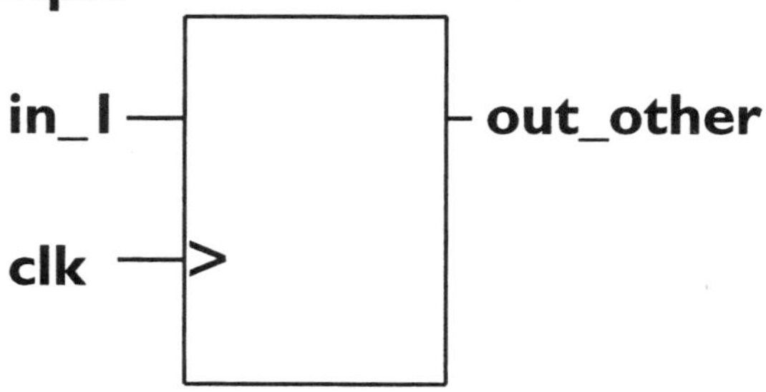


```
architecture example of rising_edge
begin
    in_out : process (clk)
    begin
        if clk : = '1' then
        out_other <= in_1 ;
    end if ;
    end process in_out ;
end example ;
```

8.9 Edge Triggering

Most synchronous devices are designed such that the changes in output occur only at a specific point in the triggering input. Typically these changes occur on either the rising edge or the falling edge of the clock pulse.

In the example in Illumination Box 8.8 the change in output occurred whenever the clock changed. If the clocking signal were of type BIT then the process took place on both the rising and the falling edge of the clock pulse. While this may model the way some devices operate it is more likely that the output changes only on the rising <u>or</u> the falling edge of the clock.

The example in Illumination Box 8.9 shows the device that appeared in the previous Box, with one modification. The change in output occurs only on the rising edge of the clock.

1. The body of the architecture consists of one **process** statement that is sensitive to the *clk* signal. Any change in the value of this signal causes the process to be executed.
2. When the process is executed the value in the signal *clk* is checked to see if it is equal to '1'.
3. If the clocking signal is '1' the value of the input signal *in_1* is assigned to the output signal *out_other.*
4. If the clocking signal is '0' no signal assignments are made.
5. When the **end process** is encountered the process is suspended until the clock again changes value.

Although the process is activated each time the clocking signal changes the assignment of the input data value to the output is done only when the clocking signal is equal to '1'. Hence the change in output occurs only on the rising edge of the clock (the process does not activate unless there is an event [change in value] of the clocking signal. If the clock is a signal of type BIT an event is either a rising edge or a falling edge of the clock signal. Checking to see if the new value of the clocking signal is '1' determines if the edge is rising or falling. A '1' shows a rising edge and a '0' shows a falling edge).

8.10 Attributes

- Specific values associated with given objects
- Certain attributes for signals, arrays and types are predefined
- Additional attributes may be defined by the user
- Format

  object_name'ATTRIBUTE_DESIGNATOR

- Example

  clock'ACTIVE

8.10 Attributes

Each item has certain characteristics which distinguish it from all other things. In human beings our distinguishing characteristics include height, weight, age, sex and other factors. These qualities are referred to as our "attributes". In general, attributes are specific values associated with given objects. Just as human beings have attributes signals, arrays and types have attributes within VHDL.

Certain attributes of signals, arrays and types have been identified and given names in VHDL. These are called "predefined" attributes. VHDL also provides a method whereby the user can identify, and give names to, other signal, array and type attributes. These are referred to as "user define attributes".

Attributes are identified by a name called an "attribute designator". A specific attribute for a specific object is accessed by using the name of the object and the desired attribute designator. For example:

object_name ' ATTRIBUTE_DESIGNATOR

The tick mark, or apostrophe, (') which appears between the object name and the attribute designator is a requirement. It indicates that the name following the object is an attribute name and not part of the object name. It also indicates that the value desired is the attribute value and not the object value.

8.10.1 Predefined Signal Attributes

- **signal'EVENT**
  returns value TRUE or FALSE if event occurred in present delta time period
- **signal'ACTIVE**
  returns value TRUE or FALSE if activity occurred in present delta time period
- **signal'STABLE (t)**
  returns a signal value TRUE or FALSE based on event in time (t) units
- **signal'QUIET (t)**
  returns a signal value TRUE or FALSE based on activity in time (t) units
- **signal'TRANSACTION**
  returns an event whenever there is an activity on the signal
- **signal'DELAYED (t)**
  returns a signal delayed (t) time units
- **signal'LAST_EVENT**
  returns amount of time since last event
- **signal'LAST_ACTIVE**
  returns amount of time since last activity
- **signal'LAST_VALUE**
  returns value equal to previous value

8.10.1 Predefined Signal Attributes

The following predefined signal attributes are part of the VHDL lexicon.

NB.: An "event" occurs when there is a change in the value of a signal. An "activity" occurs when there is either a change in the value of a signal or the value is replaced by itself. For example: a change from 'X' to '0' is both an activity and an event. A change from '1' to '1' is an activity but is not an event.

1. *signal*'EVENT - returns the value TRUE if an event has occurred on the *signal* in the present delta time period. Returns FALSE if no event has occurred.
2. *signal*'ACTIVE - returns the value TRUE if an activity has occurred on the *signal* in the present delta time period. Returns FALSE if no activity has occurred.
3. *signal*'STABLE(t) - returns a signal that has the value TRUE when an event has not occurred on the signal in the past "t" time units. The value FALSE is returned if an event has occurred in the past "t" time units. If a time period (t) is not specified this attribute returns TRUE if an event has *not* occurred in the present delta time period.
4. *signal*'QUIET(t) - returns a signal that has the value TRUE when an activity has not occurred on the signal in the past "t" time units. The value FALSE is returned if an activity has occurred in the past "t" time units. If a time period (t) is not specified this attribute returns TRUE if an activity has *not* occurred in the present delta time period.
5. *signal*'TRANSACTION - Returns a "transaction" state when an activity has occurred on the signal. This is very useful where a **process** statement has to be activated by any activity on a signal (not just by events on the signal).
6. *signal*'DELAYED(t) - Returns a signal equivalent to a *signal* delayed by "t" time units.
7. *signal*'LAST_EVENT - Returns a value equal to the amount of time that has elapsed since the last event occurred on the *signal*.
8. *signal*'LAST_ACTIVE - Returns a value equal to the amount of time that has elapsed since the last activity occurred on the *signal*.
9. *signal*'LAST_VALUE - Returns a value equal to the previous value of the *signal* (the value that the signal had before the last event which occurred on the signal).

Note that some signal attributes return signals and some return values. This is an important distinction because signals returned by attributes have attributes too. Thus certain attributes may be "strung" together e.g.

signal'DELAYED (10) 'LAST_VALUE

8.10.2 Checks with Attributes

- **Rising clock edge**
  clk'EVENT and clk = '1'
  (or) not clk'STABLE and clk = '1'

- **Falling clock edge**
  clk'EVENT and clk = '0'
  (or) not clk'STABLE and clk = '0'

- **Pulse width**
  signal'LAST_EVENT >= 10 ns

- **Stability**
  signal'STABLE (10 ns)

8.10.2 Checks with Attributes

Attributes make certain modelling operations much easier. The examples in Illumination Box 8.10.2 show some common checks which can be made using attributes.

- **Rising clock edge**

  The rising clock edge can be identified with the following check

  clk'EVENT **and** clk = '1'

  The attribute 'EVENT returns TRUE when the value of the clock has changed in the current delta time period. If the value has changed and the value is '1' then this must be the rising edge of the clock pulse. Some modelling styles use the following to identify rising edges:

  not clk'STABLE **and** clk = '1'

  Although this style may be somewhat confusing it is recommended by the VHDL-1993 LRM, viz. :

  Since the attributes S'EVENT, S'ACTIVE, S'LAST_EVENT, S'LAST_ACTIVE, and S'LAST_VALUE are functions, not signals, they cannot cause the execution of a process, even though the value returned by such a function may change dynamically. It is thus recommended that the equivalent signal-valued attributes S'STABLE and S'QUIET, or expressions involving those attributes, be used in concurrent contexts such as guard expressions or concurrent signal assignments. Similarly, function STANDARD.NOW should not be used in concurrent contexts[1].
- **Falling clock edge**

  The falling clock edge can be identified by the following check

  clk'EVENT **and** clk = '0'

  The attribute 'EVENT returns TRUE when the value of the clock has changed in the current delta time period. If the value has changed and the value is '0' then this must be the falling edge of the clock pulse.
- **Pulse width**

  The width of a signal pulse can be tested in several ways. One is:

  signal'LAST_EVENT >= 10 ns

  The attribute 'LAST_EVENT returns the time which has elapsed since the last event occurred on the signal. If this time is greater than 10 nanoseconds then the pulse is at least 10 ns wide.
- **stability**

  The stability of a signal pulse can be tested in several ways. One is:

  signal'STABLE (10 ns)

  The attribute 'STABLE returns TRUE if there has been no event on the signal in the last 10 ns.

The above do not represent the totality of checks that can be made with attributes nor are they the only way such checks can be made. They are simply examples of how some checks might be made.

8.11 Edge-triggered d_flip_flop

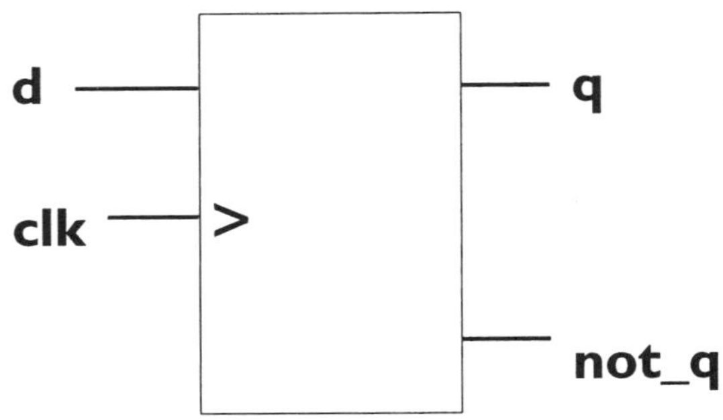


```
architecture example of d_flip_flop is
begin
    d_flop :  process (d, clk)
              begin
                  if clk'EVENT and
                      clk = '1' then
                  q <= d ;
                  end if ;
              end process d_flop ;
end example ;
```

8.11 Edge-triggered d_flip_flop

The model in Illumination Box 8.11 shows how the architecture of an edge triggered d_flip_flop might be constructed.
Consider the following specification:

> *Create a model of a rising edge triggered d_flip_flop. Input consists of a clock signal and data line. Output consists of two values, the inputted data value and the inverse of the inputted data value. Output follows input on the rising edge of the clock.*

The architecture in Illumination Box 8.11 satisfies the requirements of this specification.

1. The body of the architecture consists of one **process** statement that is sensitive to the clock signal. Any change in the value of this signal causes the process to be executed.
2. When the process is executed the condition in the **if** statement identifies whether or not there is a rising edge on the clock; (clk'EVENT **and** clk = '1' identifies a rising edge).
3. If there is a rising edge the value of the input data line is assigned to one of the output signals and the inverse is assigned to the other output signal.
4. The process then suspends until there is another clock event.

This example satisfies the conditions of the specification; however, it might give rise to several questions, such as:

- Is the clk'EVENT in the **if** statement necessary? The answer, in this case, is 'yes' since the fact that the process is being executed indicates that an event has occurred on the signal clk or the signal d. If signal d had an event the process would be executed ; but one would not wish to make the new assignment unless there had been an event on the clk (at the same time).
- Why does the data line signal appear in the process sensitivity list? The data line signal does not have to appear in the process sensitivity list. The over-riding control element is the clocking signal.

8.12 Checking Setup Time

- **Minimum interval for level prior to triggering edge for reliable entry**
- **Example**

```
architecture example of setup_check is
begin
    d_flop :    process (d, clk)
                begin
                    if clk'EVENT and clk = '1'
                        and  d'STABLE
                        (min_setup_time)
                then
                    q <= d ;
                    not_q <= not d ;
                end if ;
                end process d_flop ;
end example ;
```

8.12 Checking Setup Time

Setup time is the minimum time interval during which the input value must be stable *before* the triggering edge on the clock for reliable entry of the data to occur. An example of how this modelling problem could be applied to the d_flip_flop is shown in Illumination Box 8.12.

The only difference between this architecture and the architecture shown in the preceding Box is the addition of the line:

and d'STABLE (min_setup_time)

This line is part of the condition of the **if** statement. The value of 'STABLE is TRUE if there has not been an event on signal *d* in the previous min_setup_time period. This establishes that the input signal has been stable for the minimum time interval necessary to ensure reliable entry of the data.

Another method which achieves the same purpose is:

and d'LAST_EVENT >= min_setup_time

This line could be used to replace the previous line. 'LAST_EVENT returns the amount of time since the last event occurred on signal *d*. If that value is greater than or equal to the min_setup_time a TRUE is returned. This establishes that the input signal has been stable for the minimum time interval necessary to ensure reliable entry of the data.

There are usually several ways to approach the same test condition using attributes. Once again, if the data signal *d* were left out of the sensitivity list the clk'EVENT could be left out of the condition. Since the process would be sensitive only on the signal *clk* the modeller knows that an event has occurred on the clocking signal when the **if** statement is evaluated.

Finally, there could be a flaw in the model depicted in Illumination Box 8.12 since it does not address the problem of what to do when the data signal has not been stable for the minimum setup time. It avoids the problem by simply leaving the previous values in effect. Whether or not this is the correct action depends upon what the modeller believes the function of the device should be (or is).

8.12.1 Checking Hold Time

- **Minimum interval for level after triggering for reliable entry**

- **Example**

```
architecture example of  hold_check is
begin
process
wait until  clk = '1 and clk'LAST_VALUE = '0' ;
        wait for min_hold_time ;
        if d'STABLE (set_up_and_hold_time) then
                q <= d ;
        else
                q <= 'X' ;
        end if ;
end process ;
end example;
```

8.12.1 Checking Hold Time

Hold time is the minimum time interval during which the input value must be stable after the triggering edge on the clock for reliable entry of the data to occur. An example of how this modelling problem could be applied to an embedded d_flip_flop function is shown in Illumination Box 8.12.1. Note the use of the "double" **wait** statements:

wait until clk = '1' **and** clk'LAST_VALUE = '0' ;
wait for min_hold_time ;

The first **wait** statement waits for a rising clock edge. When a rising clock edge happens execution of the process is allowed to continue. The second **wait** statement holds the execution of the process until the minimum hold time for the data line has passed. It then allows execution to proceed.
The next line

if d'STABLE (setup_and_hold_time) **then**

returns TRUE only if the data line has been stable for the required setup and hold time. If this condition is met the **if** statement will make the assignment of the input data line to the output line

q <= d ;

If the stability conditions are not met the model assigns 'X' to the output line. The model then waits for another rising edge to occur on the clock. The expression in parentheses after d'STABLE could be a time value such as (10 ns) or an object name. Whatever is used it must be remembered that it must be a value of type TIME and it must have both an amount (10) and units (ns).

Reference

[1] *IEEE Standard VHDL Languagereference Manual* (ANSI/IEEE Std 1076-1993).

8.13 Exercises

8.13.1 A designer's assistant implemented a certain function as follows :

f := (a **and** s) **or** (b **and not** s) ;

but the designer told the assistant that the function should have been written as

if (s) **then f** := a **else** f := b ;

What was the difference between the designer's intent and the assistant's implementation?

8.13.2 Write a description for a 4-bit counter with reset and an enable for the 3-8 decoder of Chapter 3.

8.13.3 One of the classes of flip-flop is the toggled-flip-flop. The model given below has an error, what is it?

```
entity t_flip_flop is
    port (clock in BIT ;
            q : out BIT ) ;
end t_flip_flop ;
architecture switch of t_flip_flop is
begin
    process (clock)
    begin
            if clock'EVENT and clock = '1'
            then q <= not q ;
            end if ;
    end process ;
end switch ;
```

8.13.4 Create a behavioural model for an R_S flip-flop.

8.13.5 Create a model for a D-type flip_flop which matches the ac characteristics and ac setup requirements of the LS7474.

8.13.6 Write a description for the 7473 JK negative edge-triggered flip_flop whose truth table is given below :-

MODE SELECT - TRUTH TABLE

OPERATING MODE	INPUTS			OUTPUTS	
	CD_bar	J	K	Q	Q_bar
Reset (clear)	L	X	X	L	H
Toggle	H	h	h	q_bar	q
Load "0" (Reset)	H	l	h	L	H
Load "1" (Set)	H	h	l	H	L
Hold	H	l	l	q	q_bar

H, h = HIGH Voltage level
L, l = LOW Voltage level
X = Don't Care

l, h (q) = Lower case letters indicate the state of the referenced input (or output) one set-up time prior to the HIGH or LOW clock transition

Chapter 9

Conditional Assignments, Concatenation and Case

Overview

This chapter builds on the concepts of signal assignment and sequential statement control which were introduced and discussed in previous chapters. It also introduces the concept of concatenation and shows how it can be used in certain modelling situations.

In this chapter you will learn

- To use conditional signal assignment statements
- The **when** clause
- The concept of concatenation
- The **case** statement

9.1 Data Selection

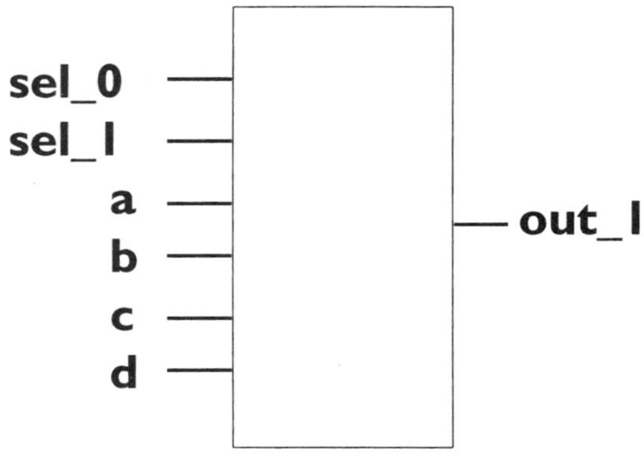


```
architecture example of multiplexer is
    begin
    one_of_four : process (sel_0, sel_1, a, b, c, d)
        begin
            if sel_0 = '0' then
                if  sel_1 = '0' then out_1 <= a ;
                else out_1 <= b ;
                end if ;
            elsif sel_0 = '1' then
                if sel_1='0' then out_1 <= c ;
                else out_1 <= d ;
                end if ;
            end if ;
    end process one_of_four ;
end example ;
```

9.1 Data Selection

A common modelling task is to create the functionality of a multiplexer. Consider the following specification:

Create a device that has six input lines and one output line. Two input lines identify which of the other four input lines are to be selected for output. These lines are called data select lines. The other four input lines are data lines. If both select lines are '0' data line a is ouputted, if '0' and '1' line b is output, if '1' and '0' line c is outputted and if they are '1' and '1' line d is output.

The model shown in Illumination Box 9.1 shows how this functionality can be achieved using **if** statement construction. Both data select signals are assumed to be of type BIT.

1. The body of the architecture consists of one **process** statement that is sensitive to all the input signals.
   Data lines are included in the sensitivity list so that the output signal updates when the data line assigned to it changes.
2. The first evaluation made during the execution of the process is the **if** condition:

   sel_0 = '0'
3. If the condition in step 2 evaluates to TRUE the nested condition is checked:

   sel_1 = '0'

   If this condition evaluates to TRUE data input line *a* is assigned to the output signal. Both statements evaluate to TRUE only when the select data lines are '0' and '0' and data line a should be output.
   If this condition evaluates to FALSE data input line *b* is assigned to the output signal. This is done with the **else**. If *sel_1* is not '0' then it must be '1' and '0' and '1' indicates that data line *b* should be ouput.
4. If the condition in step 2 evaluates to FALSE the **elsif** condition is checked:

   sel_0 = 1

   Technically, since select lines are of type BIT, the **elsif** could be replaced by a simple **else** with no condition (if *sel_0* is not '0' then it must be '1'). The **elsif** is used to clarify the actual process.
5. If the condition in step 4 evaluates to TRUE data line input *c* is assigned to the output signal. Both statements evaluate to TRUE only when the when the select data lines are '1' and '0' and data line *c* should be outputted. If this condition evaluates to FALSE data input line *d* is assigned to the output signal. This is done with the **else**. If *sel_1* is not '0' then it must be '1' and '1' and '1' indicates that data line *d* be output.

9.2 Conditional Signal Assignment

- **Form of concurrent signal assignment statement**
- **Equivalent to signal assignment using if statements**
- **Condition must evaluate to Boolean**
- **Last else must not be followed by when**
- **Format**

```
target <= waveform when condition else
          waveform ;
```

- **Example**

```
a_target <= '1' when a_source = '0' else
     'X'
```

9.2 Conditional Signal Assignment

Conditional signal assignment statements have been presented earlier in this text (Chapter 5) and used in several exercises. These statements take the form:

```
target <= waveform when condition else
          waveform ;
```

The number of **when-else** clauses in conditional signal assignment statements can be from one - to - as many as desired. However, the **when-else** conditions are evaluated in the order written. As soon as one of the conditions is true the waveform associated with it is assigned to the target and the remaining (unevaluated conditions) are skipped.

The last assignment in the statement cannot be followed by a **when-else** clause. This is to ensure that an assignment is made even if all the conditions are false.

Understanding of conditional signal assignment statements and **if** statements suggests that the conditional signal assignment statements could be replaced with **if** statements, and vice-versa. For example:

```
target <= '1' when a = '0' else
          '0' ;
```

could be replaced by:

```
process (a)
begin
   if a = '0' then
          target <= '1'
   else
          target <= '0' ;
   end if ;
end process ;
```

This comparison shows that the condition following the **when** must evaluate to a Boolean value in the same manner as the condition in an **if** statement. It also shows that the conditional signal assignment statement is a concurrent statement that is sensitive to the signals appearing in the conditions following the **when**. If a conditional signal assignment statement has several **when-else** clauses and if a different signal is used in each of the clauses than the statement is sensitive to all the signals. For example:

```
target <= '1' when a = '0' else
          '0' when b = '0' else
          'X' ;
```

This statement activates when there is an event on either signal *a* or on signal *b*. However, even if the event that activates the statement is on signal *b* the condition *b = '0'* is not checked if signal *a* is '0'.

9.2.1 Using Conditional Signal Assignment

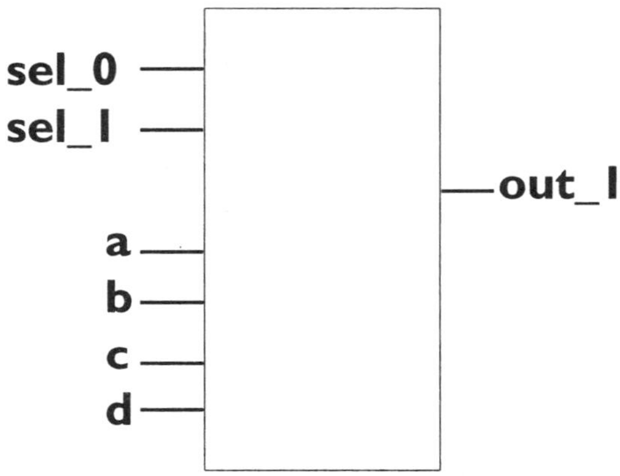


```
architecture example of multiplexer is
begin
one_of_four :
out_1 <=  a when sel_0 = '0' and sel_1 = '0' else
          b when sel_0 = '0' and sel_1 = '1' else
          c when sel_0 = '1' and sel_1 = '0' else
          d when sel_0 = '1' and sel_1 = '1' else
          'X' ;
end example ;
```

9.2.1 Using Conditional Signal Assignment

In Section 9.1 a model showing how the basic functionality of a one_of_four multiplexer could be implemented using **if** statements was presented. This model used **if** statement constructions to accomplish the following data selections:

When select line 0 is '0' and select line 1 is '0' assign data line a to output.
When select line 0 is '0' and select line 1 is '1' assign data line b to output.
When select line 0 is '1' and select line 1 is '0' assign data line c to output.
When select line 0 is '1' and select line 1 is '1' assign data line d to output.

The model in Illumination Box 9.3 shows how this functionality can be achieved using the conditional signal assignment statement. Both data select signals are assumed to be of type BIT.

1. The body of the architecture consists of a single concurrent conditional signal assignment statement.
   All signals to the right of the signal assignment operator (<=) are included in the sensitivity list of the statement. Therefore, this statement will be activated if there is an event on signals *a, b, c, d, sel_0* or *sel_1*.
2. Once the statement is activated the **when_else** conditions are evaluated in the order in which they are written. When the condition that matches the current values of the select signals is found the signal associated with the condition is assigned to the output signal.
3. All possible combinations of '1' and '0' that can occur in the select signals is accounted for in the **when_else** conditions.
   N.B.: this model would allow the use of select signals of types other than type BIT (Std_logic for example). This is possible because, if none of the **when_else** conditions evaluate as TRUE, the default assignment is 'X'. Thus, if an 'X' or 'Z' occurred on either, or both, of the select signals the output would be unknown, or 'X'.

This model shows how the functionality described by the model in Section 9.1 can be created using the conditional signal assignment statement. It is important to note that this model is easier to understand, has the added functionality of handling unknown states on the select signals, and is more efficient.

9.2.2 Selected Signal Assignment

- **Selection is based on value of included expression**
- **Format**

with expression **select**
target <= waveform **when** expression_value;

- **Example**

with a_example **select**
a_target <= waveform_1 **when** '0',
waveform_2 **when** '1';

9.2.2 Selected Signal Assignment

Another form of concurrent signal assignment is the selected signal assignment statement. In this statement selection is based on the value of an included expression. The format of this statement is:

```
with expression select
        target <= waveform when expression_value,
                  •
                  •
                  waveform when expression_value ;
```

The waveform assigned to the target signal is based upon the value of the expression after the **with**. This expression can be any object or relationship which produces a discrete value. For example, an expression like that used in an **if** statement could be used since it creates a Boolean value of either TRUE or FALSE.

```
with a <= b select
        target <= '1' when TRUE,
                  '0' when FALSE ;
```

Note that the first **when** clause is followed by a comma (,) and that the last **when** is followed by a semicolon (;). The commas indicate that another **when** follows the current clause. The semicolon marks the end of the statement. There must be at least one **when** clause in a signal assignment statement and there can be a **when** clause for every possible value that the expression can assume.

Although the above example is shown with a conditional relationship used as the expression this is not the typical usage of the selected signal assignment statement. This statement is more powerful than simply another form of the **if** statement. Typically the modeller uses an expression that produces a multitude of values and then provides an assignment for each value. As an example, suppose that the signal *foo* is type Std_logic; the following represents a simple Std_logic "inverter".

```
with foo select
        target <= 'Z' when 'X',
                  '1' when '0' ,
                  '0' when '1' ,
                  'X' when 'Z' ,
```

Selected signal assignment statements are concurrent statements; therefore they must have at least one signal that causes the statement to activate.

Note: The sensitivity list for a selected signal assignment statement consists of all signal names that appear either in the expression or to the right of the signal assignment operator.

9.3 when Clause

- **Required with selected signal assignment statements and case statements**

- **There may be any number of whens**

- **No two whens may have the same value**

- **All possible values of "expression" must be accounted for**

- **Format**

  with expression **select**
  target <= waveform **when** expression_value;

- **Example:**

```
with a_example select
a_target <= waveform_1 when '0' ,
            waveform_2 when '1' ,
            'X' when 'X' ,
            'X' when 'Z' ;
```

9.3 when Clause

The **when** clause introduced in the previous Illumination Box is required with selected signal assignment statements and **case** statements. This **when** differs from the **when** used with conditional signal assignment statements. Consider the following two formats:

Conditional Signal Assignment:

when condition **else**

Selected Signal assignment:

when expression_value,

The "condition" which appears in the conditional **when** must evaluate to a Boolean Value. The expression_value which appears in the selected **when** can be any discrete value (as long as it is part of the value set of the expression).

There must be at least one **when** clause in a signal assignment statement. At the same time the maximum number of **when** clauses is limited to the number of possible values the expression can assume. For example, if the expression in a selected signal assignment statement consisted of a single name where the signal was of Std_logic the maximum number of **when**'s would be four, one for each of the values 'X', '0', '1' and 'Z'.

An interesting aspect of the selected signal assignment statement is that all possible values of the expression must be accounted for in the **when** clause. This gives rise to an interesting question; Since the only requirement of a selected signal assignment statement is that there must be at least one **when** clause how can <u>all</u> values of the expression be accounted for without having a **when** clause for each possible value? If the expression is a Std_logic signal how could a selected signal assignment statement be written with less than four **when** clauses? Consider the following:

```
with a_example select
    target <= '1' when '0' ,
              '0' when '1' ;
```

This statement could cause an error during compilation. If *a_example* were a signal of type BIT no error would occur during compilation. All possible values that make up the value set of a BIT type object appear. If *a_example* were a signal of Std_logic an error would occur during compilation. Only two of the four possible values appear in **when** clauses. This problem could be handled as follows:

```
with a_example select
    target <= '1' when '0' ,
              '0' when '1' ,
              'X' when 'X' ,
              'X' when 'Z' ;
```

However, a better way might be to use the **others** option described overleaf.

9.3.1 others and unaffected

others

- **Indicates all unspecified elements**
- **If used, must be last item in list**
- **Example:**

```
with a_example select
a_target <= waveform_1 when '0',
            waveform_2 when '1',
            a_target when others;
```

unaffected

- **Corresponds to the null process statement**
- **Example**

```
with a_example select
a_target <= waveform_1 when '0',
            waveform_2 when '1',
            unaffected when others;
```

9.3.1 others and unaffected

The **others** option is used in selected signal assignment statements and **case** statements when the modeller does not want to provide a **when** clause for all possible values in the value set of the expression. It indicates all unspecified elements of the set. For example:

```
with a_example select
    a_target <=  '1' when '0' ,
                 '0' when '1' ,
                 a_target when others ;
```

If *a_example* is a signal of type Std_logic there are four values that must be accounted for in the selected signal assignment statement. In this example the modeller has accounted for two of the possible values. In the remaining two cases the modeller wants the value of *a_target* to remain the same. This assignment is made without discrimination as to what the values actually are; therefore the **others** option is used on the last **when** clause.

If the **others** option is used it must be the last item in the list. **others** is all inclusive. Any values remaining are included in it. In the above example the values 'X' and 'Z' are covered by the **others**. Therefore if **others** appeared before a specific value that value would never be considered.

The **unaffected** option was added for conditional assignment statements in the VHDL-1993 upgrade. **unaffected** corresponds to the **null** process statement (See Section 9.7). Using the waveform **unaffected** in a concurrent signal assignment statement means that the statement is transformed to the statement **null** in the equivalent process statement. For example:

```
with a_example select
    a_target <=  '1' when '0' ,
                 '0' when '1' ,
                 unaffected when others ;
```

causes the driver to be unchanged when the condition of *a_target* takes the values 'X' or 'Z'.

9.4 Concatenation

- **Linking two or more items into a single unit**
- **Ampersand (&) is used to indicate concatenation**
- **Items being concatenated must be of the same type**
- **Format**
  left_operand **&** right_operand
- **Example**
  signal_0 <= '0';
  signal_1 <= '1';
  signal_0 **&** signal_1 --results in "01"

9.4 Concatenation

Concatenation is the linking of two or more items into a single unit. In VHDL an ampersand (&) is used to indicate the concatenation of objects, for example, the expression:

```
left_operand & right_operand
```

indicates that the value of the object on the left side of the ampersand is to be linked to the value of the object on the right side to form a single value.

Objects that are concatenated must be of the same type, have elements of the same type (as in array elements) or have one of the objects replaced by a value from the value set of the type.

The result of the concatenation is an array whose elements are the same type as the type concatenated. Arrays are described in Chapter 10.

For example:

```
signal_0 <= '0' ;
signal_1 <= '1' ;

signal_0 & signal_1      -- results in "01"
```

If *signal_0* and *signal_1* are Std_logic objects the result is an array whose elements are of type Std_logic.

Arrays may be thought of as collections of values, therefore the concatenation of scalar values creates an array. If the objects *signal_0* and *signal _1* in the example are Std_logic objects the result of their concatenation is an array of two Std_logic_type elements.

The example shows the concatenation of Std_logic scalar objects. In actuality this is a special case. Concatenation in VHDL is normally thought of as the joining together of one-dimensional arrays whose elements are of the same type. However, a scalar value can be thought of as an element in a one-dimensional array of length one and type equal to the scalar type. Because the VHDL system is capable of resolving a value to a value set (by overloading) a scalar object, or an array, can be concatenated with a value from the type's value set.

The only drawback to concatenation in VHDL is that VHDL does not have a mechanism for deducing the kind and size of the array that is created by the concatenation process. Therefore, the concatenated object must be assigned to an array object of the correct type and size (length) *or* VHDL must be told what it has created, using a typemark. The use of the typemark is discussed overleaf.

9.4.1 Typemarks

Concurrent one-of-four

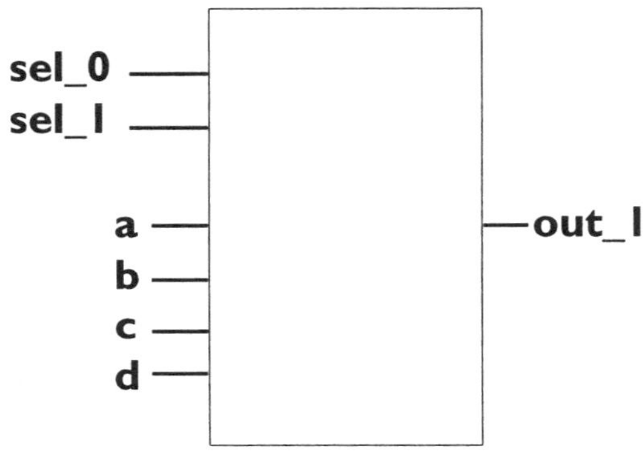


```
architecture example of multiplexer is
type cat_ray is array (0 to 1) of Std_logic;
begin
    one_of_four:
    with cat_ray'(sel_0 & sel_1) select
        out_1 <= a when "00",
                 b when "01",
                 c when "10",
                 d when "11",
                 'X' when others;
    end example;
```

9.4.1 Typemarks

Although **types** will be discussed in detail in Chapter 12 it is useful to introduce them here in respect of their relationship to concatenation.

The *one_of_four* multiplexer functionality shown in Illumination Boxes 9.1 and 9.2.1 can also be modelled using a selected signal assignment statement and concatenation. If the values of the select lines can be concatenated into a single value, this new value can be used to select which data line is outputted. However, one aspect of concatenation in the "expression" of selected signal assignment statements and **case** statements requires the introduction of this new topic at this stage.

The array which is created cannot be checked against the **when** expression_value unless the type of the elements and the length of the array that is created are known. This can be handled by creating a "qualified" expression.

Qualifying an expression is performed in order to identify the exact outcome of the expression to the system. In the case of concatenation within a selected signal assignment statement VHDL is unsure of the type of elements and the length of the array that is created. This ambiguity can be eliminated by qualifying the expression.

Qualification of an expression is performed by enclosing the expression in parentheses and preceding the parenthetical expression with the name of an unambiguous **type** and a tick mark ('). The format is:

type'(expression)

In the example in Illumination Box 9.4.1 a type called *cat_ray* is declared. This type is an array with two elements whose type is Std_logic. It is used to qualify the expression in the selected assignment statement:

cat_ray '(sel_0 & sel_1)

This tells the system that the value created by the concatenation is an array of length two whose elements may be 'X', '0', '1' or 'Z'.

The model in Illumination Box 9.4.1 shows how the one_of_four functionality can be achieved using selected signal assignment statement construction.

1. The body of the architecture consists of a selected signal assignment statement.
2. The first evaluation made during the execution of this statement is:

   sel_0 & sel_1
3. The qualification of the expression indicates that the result of the evaluation is an array of length two whose elements are Std_logic types.
4. The result is compared with the expression_values which follow the **when**s in the statement to check for a match.

9.5 Sequential Selection

- **when-else and with-when structures cannot be used in sequentially processed model areas**

- **In sequential areas if statements and case statements are used to select statements to perform**

- **Example**

  concurrent:

  ```
  with in_put select
      out_put <= '0' when '1',
                 '1' when '0';
  ```

  sequential:

  ```
  case in_put is
      when '1' => out_put <= '0';
      when '0' => out_put <= '1';
  end case;
  ```

9.5 Sequential Selection

*Conditional signal assignment statements (***when-else***) and selected signal assignment statements (***with-when***) cannot be used in areas of the model where statements are processed sequentially. Both of these statements are forms of concurrent signal assignment.*

However, there is also a need to select which assignments are to be executed within sequentially processed statement areas. Earlier in this chapter the relationship between conditional signal assignment statements and **if** statements was discussed. For example:

```
if a = b then
    target <= c ;
else
    target <= 'X' ;
end if ;
```

is functionally similar to:

```
target <= c when a = b else
          'X' ;
```

This shows how similar actions are achieved in both concurrently processed areas and sequentially processed areas.

The selected signal assignment statement also has a sequential equivalent. It is the **case** statement. The similarities are apparent by inspection:

Selected Signal Assignment:

```
with in_put select
    out_put <=   '0' when '1' ,
                 '1' when '0' ;
```

case Statement:

```
case in_put is
    when '1' => out_put <= '0' ;
    when '0' => out_put <= '1' ;
end case ;
```

In both cases it is apparent that the value of *in_put* is used to select what is going to be assigned to *out_put*.

NB. For a definition of the symbol (=>) see Section 11.2.8 (Named Association).

9.6 case Statement

- **Controls execution of one or more sequential statements**

- **Format**

case expression **is**
when exprn_vale => seq_statements;
end case;

- **Example**

```
architecture example of case_statement is
begin
        in_vert: process (in_put)
                begin
                case in_put is
                        when '1' => out_put <= '0';
                        when '0' => out_put <= '1';
                                    def_con <= 'X';
                end case;
```

9.6 case Statement

The **case** statement has many of the same properties as the selected signal assignment statement.

1. The choice of which statements are to be executed is based upon the value of an expression.
2. The expression can be any object or comparison that produces a discrete value.
3. There must be at least one **when** clause in the **case** statement and the maximum number of **when** clauses is limited to the number of values in the value set of the expression.
4. All values in the expression's value set must be accounted for in the **when** clauses.
   This does not mean that there must be a **when** clause for each value. Values can be lumped together with the **others** option.
   As might be expected, there are differences between the two statement types:

   case *statements appear within sequential code areas of the model.*

   If the **case** statement is contained within a **process** statement the **case** statement is usually executed when the process statement is evaluated.

   case *statements are not limited to the assignment of a waveform to a signal*

The selection in a **case** statement identifies a sequential statement or *series* of sequential statements to execute.
The **case** statement format is:

```
case expression is
    when expression_value => (sequential_statements ; )
                •
                •
    when expression_value => (sequential_statements ;)
end case ;
```

The (sequential_statements) is exactly what it indicates; any sequential statement or series of sequential statements. In many ways this ability to do more than simply assign a waveform to a signal makes the **case** statement more powerful than the selected signal assignment statement.

9.7 null

- **Sequential statement indicating that no action occurs**
- **An action is required for all "expression" values in case statement**
- **Used for "expression" values which require no action**
- **Example**

```
case opcode  is
        when '0' => out_1 <= in_1;
        when '1' => out_1 <= in_0;
        when others => null;
end case;
```

9.7 null

The rule that all of the values of the expression must be accounted for in the **case** statement presents an interesting dilemma. What happens if there are some expression values with which the modeller is not concerned?

Suppose that the expression in a **case** statement consisted of a Std_logic signal:

```
case opcode is
```

There are four values which opcode can assume; 'X', '0', '1' and 'Z'. Now suppose that the modeller is interested only in those cases when opcode is '0' or '1'. We are aware that all other values can be lumped into the **others** category:

```
case opcode is
    when '0' => out_1 <= in_1 ;
    when '1' => out_1 <= in_0 ;
    when others => ? ;
end case ;
```

The modeller knows that the **others** is necessary to account for 'X' and 'Z'. However, the modeller does not want to do anything when the value of opcode is either 'X' or'Z'. How does the modeller handle this dilemma? There are values for which nothing is to be done, but an action is required for all values of the expression in a **case** statement (e.g. **when** this **do** this).

The **null** statement is a sequential statement which can be used to indicate that no action is to take place. The above example could, therefore, be completed as follows:

```
case opcode is
    when '0' => out_1 <= in_1 ;
    when '1' => out_1 <= in_0 ;
    when others => null ;
end case ;
```

9.8 Sequential one_of_four

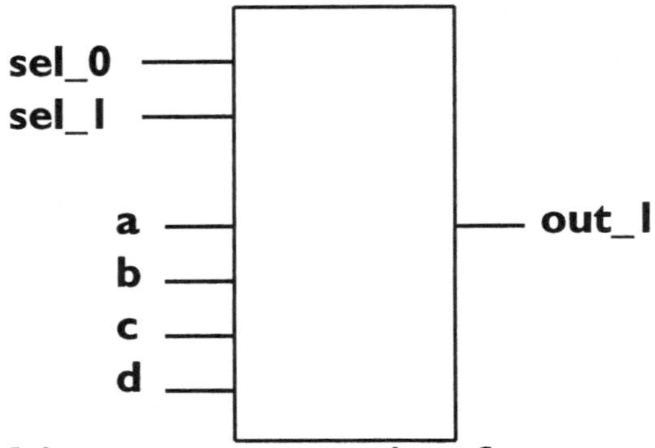


```
architecture example of case_mux is
type cat_ray is array (0 to 1) of Std_logic;
begin
one_of_four:
    process (sel_0,sel_1, a, b, c, d)
        begin
        case cat_ray'(sel_0 & sel_1) is
            when "00" => out_1 <= a ;
            when "01" => out_1 <= b ;
            when "10" => out_1 <= c ;
            when "11" => out_1 <= d ;
            when others => null ;
        end case;
    end process one_of_four;
end example;
```

9.8 Sequential one_of_four

The one_of_four multiplexer shown in Illumination Boxes 9.1, 9.2.1 and 9.4.1 can also be modelled using a **case** statement and concatenation. This is to be expected considering the fact that a selected signal assignment statement is simply a concurrent form of a **case** statement and the fact that the one_of_four functionality could be shown with a selected signal assignment statement.

Like the selected signal assignment statement the expression in the **case** statement must be "qualified" if it involves the concatenation of values.

The model in Illumination Box 9.8 show how the one_of_four functionality can be achieved using a **case** statement and concatenation.

1. The body of the architecture consists of a process statement which is sensitive to the input signals *sel_0, sel_1, a, b, c* and *d.*
   Remember that the selected assignment statement was sensitive to all signals in its expression and to the right of the signal assignment operator. The **case** statement is a sequential statement. It must appear in a sequential area (for now, within a **process** statement). It executes whenever the process statement executes. Therefore, to model the functionality all the input signals must appear in the process sensitivity list.
2. The first evaluation made during execution of the process is:
   sel_0 & sel_1
3. The qualification of the expression indicates that the result of the evaluation is an array of length two whose elements are Std_logic types (See Chapter 10).
4. The result is compared against the expression_values that follow the **when**s in the **case** statement. If a match is found the statements associated with that **when** are executed.
   In this example only a single signal assignment operator follows each **when**. The function of this statement is to assign the appropriate input value to the output signal.
   If none of the values in the **when** match the result no action is taken. This is done by the **when** clause:
   when others => **null**;
   The **others** traps all values which have 'X' or 'Z' in them. The **null** indicates no action.

9.9 one_bit adder

(an exercise in concatenation)

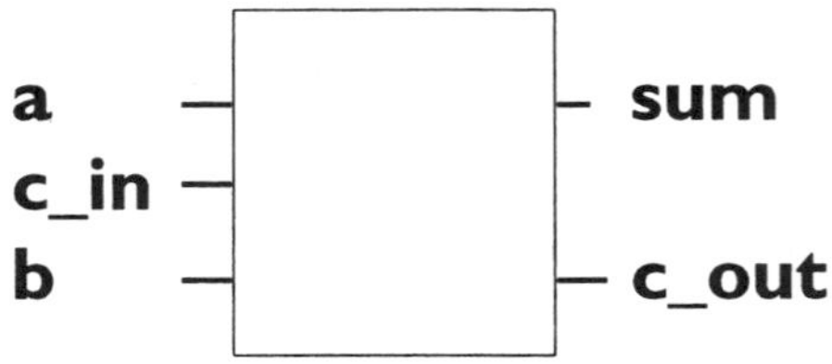


```
architecture behav of one_bit_adder is
        signal total : INTEGER RANGE 0 TO 4;
        type example is array (0 to 2) of Std_logic;
begin
majority: with example'(a & c_in & b) select
        total <=    0 when "000",
                    1 when "100" | "010" | "001",
                    2 when "101" | "011" | "110",
                    3 when "111",
                    4 when others;
adder:      process (total)
            begin
                case total is
                  when 0 => sum <= '0'; c_out <='0';
                  when 1 => sum <= '1'; c_out <='0';
                  when 2 => sum <= '0'; c_out <='1';
                  when 3 => sum <= '1'; c_out <='1';
                  when others => sum <= 'X';
                                 c_out <= 'X';
                end case ;
            end process adder ;
end behav ;
```

9.9 An Exercise in Concatenation

By now it has become apparent that there are many ways to model a single unit of functionality. The model in Illumination Box 9.9 shows how the simple one_bit_adder (xor_gate) which was presented in Chapter 5 can be expressed using selected signal assignments and **case** statements.

Note: This is not meant to be an example of how to model a one_bit_adder; it is simply a mechanism for showing several interesting techniques.

1. The body of the architecture consists of two concurrent statements; a selected assignment statement and a process statement.
   The selected signal assignment statement is labelled majority. It is sensitive to the input signals *a, b* and *c_in.*
   The process statement is labelled adder. It is sensitive to the signal total. This signal is an internal signal. Values are assigned to it by the selected signal statement.
2. Whenever one or more of the input signals change the selected signal assignment statement is activated.
   The evaluation of the expression *a & c_in & b* produces an array of length three whose elements are Std_logic types. The expression is qualified so that it can be compared with the expression values.
3. Comparison of the values causes a 0, 1, 2, 3 or 4 to be assigned to the internal signal total. Notice that these are integer numbers and that the integer value is equal to the number of 1s which appear on the three input signals.
4. Once the selected signal assignment statement has finished executing, the value assigned to total may cause the process statement to activate (this means that there may be a second iteration).
5. The **case** statement uses the value of total to determine which signal assignment to make to the output signals *sum* and *c_out.*
   The outputs are consistent with a truth table of the inputs. When the sum of the inputs is zero the outputs are zero. When the sum is one the output sum is one and the carry is zero; and so on. Any time one of the inputs is unknown, both the outputs are set to unknown.

The model shows most of the topics presented in this chapter. Selected signal assignment statements, **case** statements, concatenation, qualified expressions and the **others** option.

Notice the use of the "pipeline" character in the selected signal assignment statement. This character says that the values on either side of it are to be treated as the same unit. This does not mean that they are the "same". It simply means that, in this usage, they have equivalent meaning.

9.10 Exercises

9.10.1 The following code is for a 4-bit and_gate with enable. What error would a compiler indicate and how could it be corrected?

```
entity \4_bit_enabled_gate\ is
    port (a : in BIT_VECTOR ( 3 downto 0) ;
          enable : in BIT ;
          b : out BIT_VECTOR (3 downto 0)) ;
end \4_bit_enabled_gate\ ;
architecture wrong of \4_bit_enabled_gate\ is
begin
    b <= a and enable ;
end wrong ;
```

9.10.2 Create a model for the 4-bit comparator in the 3-8 decoder of Chapter 3.

9.10.3 Write code for an 8-bit 2 to 1 multiplexer.

9.10.4 Write code for a 16-bit adder/subtractor.

9.10.5 Produce a VHDL file for an 8-bit register.

9.10.6 Design an 8-bit register which outputs positive numbers only and rejects negative numbers. Assume the existence of a sign bit.

9.10.7 Write behavioural code for a one_of_four multiplexer using nested **case** statements.

9.10.8 Create a model of a simple priority decimal to binary encoder. The input lines will represent the decimal numbers 0 to 7 and the 'priority' means that the encoder will produce a binary output corresponding to the largest decimal digit appearing on the inputs. For example, if the two input lines corresponding to decimal 4 and decimal 2 are high, the output on the three binary lines should be 1, 0, and 0, which represents '4' in binary; if all inputs are 0 the output should also be set to zero.
NB. Particular consideration should be given to the use of **case** and concatenation among the many possible solutions.

9.10.9 Write VHDL descriptions for the 3-8 decoder of Chapter 3 using:
i) Selected Signal Assignments
ii) Case Statements
iii) Concatenation
For this version assume that 'Z' has the same effect as '1' for the IEEE Library values. What type of assignment was used for Structure_2?

Chapter 10

Arrays, Loops and Assert Statements

Overview

The topic of arrays is re-introduced and discussed extensively in this chapter. Because arrays lend themselves to iterative processes loops are introduced and developed. A method for checking conditions and issuing reports (the ASSERT statement) is also presented in this chapter

In this chapter you will learn

- Formal methods for declaring and using arrays
- Iterative and conditional control of sequential code
- The structure and use of the **assert** statement

10.1 Array Input one_bit_adder

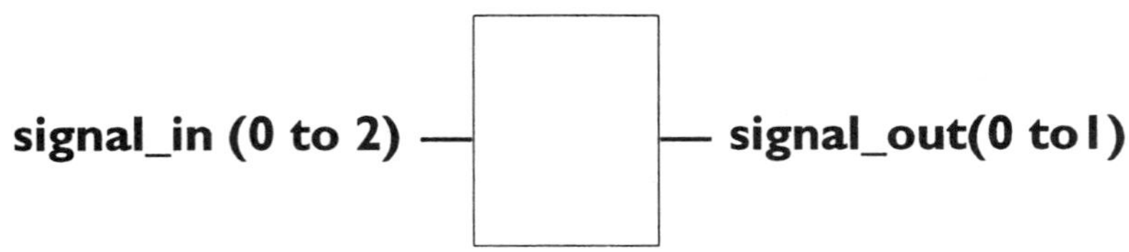


```
entity adder is
        port (signal_in: in BIT_VECTOR (0 to 2);
                signal_out:out BIT_VECTOR (0 to 1);
end adder ;
architecture array_example of adder is
begin
        add_up: process (signal_in)
                begin
                        case signal_in is
                                when "000" =>
                                                signal_out <= "00";
                                when "001" | "010" | "100" =>
                                                signal_out <= "10";
                                when "011" | "101" | "110" =>
                                                signal_out <= "01";
                                when "111" =>
                                                signal_out <= "11";
                        end case;
                end process add_up;
end array_example;
```

10.1 Array Input one_bit_adder

A model of a simple one_bit_adder using a selected signal assignment and a **case** statement was presented and discussed in Chapter 9 (Section 9.9). The function of the selected signal assignment statement was to convert the three individual input signals into a single value. This was done by first concatenating the three values and then assigning an integer value equal to the number of 'ones' ('1') in the array to an intermediate signal. This value was then used in a **case** statement to make the correct value assignments to the output signals.

The obvious question arising from the above is "Why not simply perform the concatenation in the **case** statement and eliminate the selected signal assignment statement?". True, the selected signal assignment statement could have been eliminated in this way but the model in Section 9.9 was developed in order to demonstrate the use of both statements. There are several ways in which the model may be improved. The example in Illumination Box 10.1 shows an evolution of the model for a one_bit_adder.

Concatenation produces an array of values. However, suppose that, instead of having to create the array, the values were input as an array. This would eliminate the need to concatenate the signals and would allow the input to be used directly in a **case** statement. Consider the model in the Illumination Box 10.1:

1. Input and output are declared to be bit vectors.
   Vector is the term used to refer to a one dimensional array. The generic term "array" is used throughout this book to refer to arrays of any dimension.
   The **type** BIT_VECTOR is defined in the VHDL package STANDARD. It is used in this example in order to avoid having to deal with 'X' and 'Z' values.
2. The body of the architecture is made up of one process statement which contains a **case** statement. This **process** statement is sensitive to the input signal signal_in. When any element in the array changes, the process is activated.
3. The expression in the **case** statement is the input signal. Therefore the values which appear in the **when**s must be the values which signal_in can assume. Since these values are array values they must appear in double quotes (see Section 5.3.3).
   Note the use of the vertical lines (pipeline) between some values in the **when**s. These indicate that such values are to be treated as equivalent.
4. Output is also treated as an array. When the input value is identified the appropriate output is assigned to the output signal signal_out.

10.2 Arrays

- **Collections of values all belonging to the same object**
- **Individual values stored within separate elements within object**
- **Individual elements identified by an index value**
- **Set up requires an array type declaration and an object declaration of that type**
- **Constrained and unconstrained array declarations, but only constrained objects**

- **Type Declaration Format**

type name **is array** (index_range) **of** element_type ;

- **Examples:**

  type array_type **is array** (0 **to** 10) **of BIT ;**
  type BIT_VECTOR is array
  (NATURAL RANGE <>) of BIT ;
  signal foo: array_type ;
  signal boo: **BIT_VECTOR** (0 **to** 3) ;

Introduction to VHDL

10.2 Arrays

Earlier in this book arrays were described as collections of values. Each of these values belongs to the same object. Individual values in an array are stored in separate elements. These elements are identified by index values. Because arrays are collections of values they are referred to as "composite" types.

In some ways an array resembles an apartment building. The building has a single name (address). The apartments within the building are numbered so that the individuals who live therein can be accessed. An array object is an "apartment building" for values. Individual values are placed in elements in the array (apartments). These elements are identified by index values (apartment numbers)..

The index values of an array must be distinct. Integer numbers and enumerated type values are examples of distinct values. Such values are often called discrete values.

There are two kinds of array types:

1. Constrained array types

   Constrained array types have limits placed on the number of elements in their type declaration. For example:

   type example **is array** (0 to 10) **of** BIT ;

   This is a type declaration for an array type called example. Arrays of this type are limited to eleven elements which would be numbered 0, 1, 2, 3, 4, 5, 6, 7, 8, 9, 10.

2. Unconstrained array types

   Unconstrained array types have no limits placed on the number of elements in their type declaration.

   NB. There is always an implicit limit in the type declaration on the number of elements because the size of the array has to be stated in terms of a discrete type. Each discrete type has a range associated with it. For example:

   type BIT_VECTOR **is array** (NATURAL RANGE <>) **of** BIT;

   This is the type declaration which appears in the VHDL package STANDARD for the type BIT_VECTOR. It is an unconstrained array. The size of the array (NATURAL_RANGE <>) is left open. Array objects declared to be of this type can have the number of elements stated when the object is declared. The only limit on the number of elements is that they must be within the range of the type NATURAL. Even though an array type can be declared as unconstrained, all array objects must be constrained when they are declared.

This discussion applies only to the declaration of array types. Discussion of the declaration of array objects is presented in Chapter 12.

10.2.1 Predefined Array Attributes

- **Return information about index values**

- **Formats**

array_name'LEFT **array_name'RIGHT**
array_name'HIGH **array_name'LOW**
array_name'RANGE **array_name'REVERSE_RANGE**
array_name'LENGTH

- **Examples**

signal ray : **BIT_VECTOR** (0 **to** 7) ;

ray'LEFT	**returns 0**
ray'RIGHT	**returns 7**
ray'HIGH	**returns 7**
ray'LOW	**returns 0**
ray'RANGE	**returns 0 to 7**
ray'REVERSE_RANGE	**returns 7 downto 0**
ray'LENGTH	**returns 8**

10.2.1 Predefined Array Attributes

Just as there are attributes associated with scalar objects, there are attributes associated with array objects. These attributes return information about the index values of the arrays.

Consider the following object declaration:

signal ray : BIT_VECTOR (0 **to** 7) ;

This statement declares an array object called ray. The BIT_VECTOR (0 **to** 7) indicates that this array has eight elements which can hold values of type BIT. The type BIT_VECTOR is a predefined unconstrained array type composed of elements of type BIT. Since it is unconstrained, and since array objects must be constrained, the range (0 **to** 7) indicates the number of elements in the array.

There are seven array attributes which can be applied to array objects like the one in this example:

1. array_name'LEFT - Returns the left most index value in the range of the array. ray'LEFT returns 0.

2. array_name'RIGHT - Returns the right most index value in the range of the array. ray'RIGHT returns 7.

3. array_name'HIGH - Returns the largest index value in the range of the array. ray'HIGH would return 7.

4. array_name'LOW - Returns the smallest index value in the range of the array. ray'LOW would return 0.

5. array_name'RANGE - Returns the entire range of the array. ray'RANGE returns 0 to 7.

6. array_name'REVERSE_RANGE - Returns the reverse range of the array. ray'REVERSE_RANGE returns 7 downto 0

7. array_name'LENGTH - Returns the number of elements in the array. ray'LENGTH would return 8.

Although in this example 'RIGHT and 'HIGH return the same value, this is not always the case. 'RIGHT and 'HIGH return the same value when the range of the array is written as an ascending range (some_value **to** some_larger_value). They will return different values when the range is written as a descending range (some_value **downto** some_smaller_value). 'HIGH always returns the maximum index value independent of the direction indicated in the range.

10.3 Loops

- **Sequences of statements which are executed repeatedly**

- **Types of loops**

  - **for**
  - **while**
  - **loop-exit construct**

- **Format**

```
[loop_label : ]
    iteration_scheme
        loop
        sequence_of_statements
        end loop [loop_label] ;
```

10.3 Loops

Loops are repetitive actions. In VHDL a loop is a sequence of statements which are executed repeatedly each time it is invoked. Loops in VHDL are a way of controlling code in sequential areas of the model. There are three formal methods of creating loops:

1. **for** loops
   These loops execute once for each value in the parameter range of the **for** expression. The loop executes a set number of times each time it is invoked.
2. **while** loops
   The number of times loops of this type may execute is controlled from within the loops themselves. The loop is exited from when the Boolean condition stated in the **while** expression becomes FALSE.
   NB. Some factor within the loop must affect the condition or the result could be an 'infinite' loop.
3. **loop - exit** constructs
   The **loop - exit** construct has no statement condition for termination nor does it have a fixed number of times to be repeated. The decision to leave loops of this kind has to be made within the loop itself.

Note that loops are the only sequential statements that can have labels.

The general format for loops is:

```
[loop_label : ]
    iteration_scheme
    loop
            sequence_of_statements
    end loop [ loop_label ] ;
```

The loop label is optional. However, if it appears at the beginning of a loop it must also appear after the **end loop** statement.

The iteration_scheme is either a **for** expression or a **while** expression. The **loop - exit** construct does not have an iteration scheme. It simply begins with the key word **loop**.

loop identifies the beginning of the loop. **end loop** identifies the end of the loop. All the statements which appear between **loop** and **end loop** are said to be in the sequence of statements which is to be repeated.

10.3.1 for loop

- **Statements are executed once for each value within the range of the loop parameter**
- **Loop parameter is declared implicitly and may not be modified within the loop or used outside it**

- **Format**

```
[label :] for loop_parameter in discrete_range
          loop
               --sequential_statements
          end loop [label] ;
```

- **Example**

```
process (signal_a)
begin
lbl: for index in 0 to 7
     loop
       ray_out (index) <= ray_in (index);
     end loop lbl;
end process ;
```

10.3.1 for loop

for loops are so-named because the iteration_scheme in these statements begin with the word **for**. Statements within these loops are executed once for each value in the parameter range of the **for** expression.

```
[loop_label :]
    for loop_parameter in discrete_range
    loop
            -- sequence_of_statements
end loop [loop_label];
```

The loop parameter in the **for** expression is implicitly declared. An object declaration does not have to be created for this parameter; however; the name given to the parameter must follow the standard VHDL object naming rules.

Although the value of the loop parameter may not be modified either inside, or outside, the loop it can be referenced within the loop.

A loop parameter cannot be referenced outside the loop within which it resides.

Consider the following example:

```
lbl : for index in 0 to 7
loop
    ray_out (index) <= ray_in (index) ;
end loop lbl ;
```

This example would transfer the elements of an array called ray_in to an array called ray_out. Values would be transferred between corresponding elements, i.e., the value in element 0 of the input array would be transferred to element 0 of the output array.

1. lbl: **for** index **in** 0 **to** 7
   The loop is lbl. The loop parameter is index. The range of the object index is from integer 0 to integer 7. There are eight discrete values in this range; therefore, it would be expected that the statements in the loop would be executed eight times.
2. **loop** and **end loop** lbl ;
   Mark the beginning and the end of the statements which are to be repeated.
3. ray_out (index) <= ray_in (index) ;
   This statement transfers the values between the two arrays. Note the use of the **for** parameter (index). The first time through the loop the value of (index) is 0. Hence the value in *ray_in (0)* is transferred to

10.3.2 while loop

- **Execution of statements within loop is controlled by Boolean condition**

- **Condition is evaluated before each repetition of loop**

- **Format**

```
while boolean_expression
    loop
        -- sequential_statements
    end loop ;
```

- **Example**

```
p1 : process(signal_a)
    variable index : INTEGER := 0 ;
    begin
    from_in_to_out :
        while index < 8 loop
          ray_out (index) <= ray_in(index) ;
          index := index + 1 ;
        end loop from_in_to_out ;
    end process p1 ;
```

ray_out(0). The last time through the loop the value of (index) will be 7 and the value of *ray_in (7)* will be transferred to *ray_out (7)*. This is a valid use of the parameter since it is referenced only and is not modified.

The restriction on the index values is that they must be discrete values. Integer numbers satisfy this requirement and are, typically, so used. However, any discrete value set can be used, for example, *a* **to** *f*.

10.3.2 while loop

while loops are so-named because the iteration_scheme in these statements begin with the word **while**. Statements within these loops are executed so long as the Boolean condition in the **while** expression evaluates as TRUE.

```
[loop_label : ]
    while boolean_expression
    loop
            -- sequence_of_statements
    end loop [loop_label] ;
```

The number of times **while** loops execute is controlled from within the loops themselves. The loop is exited when the condition stated in the **while** expression evaluates as FALSE. This condition is evaluated before each repetition of the loop. Therefore, if the condition is FALSE when the loop is first encountered the statements within the loop are not executed. If the condition is TRUE the statements within the loop are executed. So, unless something in these statements affects the Boolean condition, an infinite loop exists.

The condition can be anything which evaluates to a Boolean result (i.e. a relational comparison which results in a Boolean value). Consider the following example:

```
p1 :    process (signal_a)
                variable index :  INTEGER := 0 ;
        begin
                from_in_to_out :
                while index <   8 loop
                        ray_out (index) <= ray_in (index) ;
                        index := index + 1 ;
                end loop  from_in_to_out ;
        end process p1 ;
```

This example accomplishes exactly the same function as does the **for** example in Illumination Box 10.3.1; but with some notable differences.

1. **while** loops do not implicitly define parameters. If an object is going to be used as a "counter" in the loop it must be declared. Note that the variable index is declared in the process declarative area. There is no

10.3.3 loop-exit Construct

- **Does not contain an iteration scheme specifying how loop terminates**

- **A decision to leave loop must be made from within the loop**

- **Can be terminated from within by:**

  - **exit**
  - **next**

- **Format**

```
loop
    -- sequential statements
    -- including decision_to_quit
end loop ;
```

declarative area in the loop. Also note that index is modified within the loop. Since this variable is a declared object it may be modified as desired.

2. The Boolean condition in this **while** loop is index < 8. So long as the value of index is less than eight the loop will be repeated. Since the initial value of index is 0, and since 1 is added to index each time through the loop, the loop will be repeated eight times. If the statement

   index := index + 1 ;

   were to be removed the loop would repeat indefinitely.

10.3.3 loop - exit Construct

The **loop - exit** construct is so-named because it does not contain an iteration scheme specifying how the loop terminates. Instead, a decision to leave the loop must be made from within the loop.

The format of the **loop - exit** construct is:

```
[loop_label :]
  loop
        -- sequence of statements
        -- including decision_to_quit
  end loop [loop_label] ;
```

Without an explicit decision to quit, the **loop - exit** construct creates an infinite loop. Termination of the loop can be brought about using either of two statements:

1. **exit**
   Used to break out of loops.
2. **next**
   Causes completion of one of the iterations of a loop; or may be used to leave a "labelled" loop.

Every **loop - exit** construct must have at least one of these two statements in its body to avoid entering an infinite loop condition.

10.4 exit Statement

- **Used to break out of loops**
- **Only occurs within loops**
- **Format**

```
exit ;
exit loop_label ;
exit when boolean_expression ;
exit loop_label when boolean_expression ;
```

- **Example**

```
show_loops :  process (s)
                  variable sum, cnt : INTEGER := 0 ;
    begin
        sum := 0 ; cnt := 0 ;
        first :     loop
                        cnt := cnt + 1 ;
                        sum := sum + cnt ;
                        exit when sum > 100 ;
                    end loop first ;
        second :    loop
                        cnt := cnt + 1 ;
                        sum := sum + cnt ;
                        if sum > 100 then exit ;
                        end if ;
                    end loop second ;
    end process show_loops ;
```

10.4 exit Statement

The **exit** statement is used to terminate loops. When it is executed it causes immediate cessation of the statements in the loop. Any statements following **exit** are not executed. Evaluation skips to the statement immediately following the **end loop** statement.

exit *may appear only in loops. It cannot be used outside the loop structure.*

Although **exit** is presented here as being used in the **loop - exit** construct it can be used in both **for** loops and **while** loops.

The **exit** statement has four formats:

1. **exit;**
   This form of the **exit** statement is unqualified. When it is encountered, the loop in which it is immediately contained is terminated.
2. **exit** loop_label;
   This form is slightly qualified by giving the label of the loop which is to be terminated. If the loop_label which follows **exit** is the label of the immediate loop there is no difference between this form and the previous form. However, it is possible to have "nested" loops (loops within loops). In these cases it is possible to state the label of a loop which is not the immediately enclosing loop. When this happens both the immediate loop and the loop containing the immediate loop are terminated. Since there could be multiple layers of loops it is important to realise that all loops out to, and including, the loop label which appears in the **exit** statement are terminated.
3. **exit when** Boolean_expression;
   It is possible to qualify the **exit** statement by stating when the exit is to take place. The Boolean expression can be either an object or a relational expression which evaluates to a Boolean value. A **true** value causes termination.
4. **exit** loop_label **when** Boolean_expression;
   This form shows both types of qualification; which loop and when. All of the conditions of termination apply in this case.

An alternative to using **when** qualification with **exit** would be to place the **exit** under the control of an **if** statement. For example, the following two statements are equivalent:

```
exit when sum > 100;
if sum > 100 then exit;
```

10.5 next Statement

- **Causes execution to jump to the beginning of the loop and start the next iteration**

- **Format**

```
next ;
next loop_label ;
next when boolean_expression ;
next loop_label when boolean_expression ;
```

- **Example**

```
nxt_xmpl : process (s)
             variable temp, j : INTEGER:= 0 ;
           begin
           temp := 0 ; j := 0 ;
           a_loop :    for i in 0 to 7 loop
                           j := j + 1 ;
                           if i > 5 then
                           next a_loop ;
                           end if ;
                           temp := temp + 1 ;
                       end loop a_loop ;
           end process nxt_xmpl :
```

10.5 next Statement

The **next** statement causes completion of one of the iterations of an enclosing loop. This may, or may not, cause termination of a loop.

next causes the next cycle of the loop to begin. Execution of statements begins at the first statement following the loop statement. Any statements which occur after the **next** and before the **end** loop are not evaluated. In those cases where **next** refers to an outer loop, the inner loop is terminated and the next iteration of the out loop is begun. **next** appears only in loops. It cannot be used outside the loop structure. In most cases it is used with for loops and **while** loops rather than **loop-exit** constructs.

The **next** statement has four formats:

1. **next**;
   When this form is encountered evaluations of statements begins at the first statement following **loop**. In the case of **for** loops, the loop parameter is incremented to indicate that the next iteration has begun. In all cases the statements following **next** are not evaluated once the **next** statement is executed. It should be noted that a misused **next** can easily create an infinite loop in **while** loops and **loop-exit** constructs.
2. **next** loop-label;
   The loop_label specifies the loop upon which the next iteration is to begin. In this form the **next** statement can cause the termination of a loop. Consider two nested **for** loops: the outer loop is labelled *loop_a* and the inner loop *loop_b*. In *loop_b* is the statement **next** *loop_a*. When this statement is executed the next iteration of the outer loop is begun. To do this requires leaving the loop and effectively terminating its execution.
3. **next when** Boolean_expression;
   The **when** condition qualifies the **next** statement by placing a condition which must be met before the **next** can be executed. The Boolean expression can be either an object or a relational expression which evaluates to a Boolean value. A TRUE value causes the next iteration to begin.
4. **next** loop_label **when** Boolean_expression;
   This form shows both types of qualification, which loop and when. All of the conditions stated in 2 and 3 apply in this case.

10.6 Four-to-sixteen Decoder

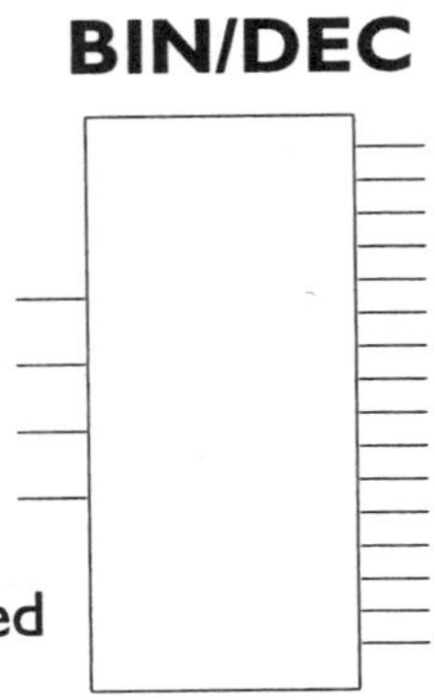

in_array
(0 to 3)
(an array of integers whose values are limited to 0 to 1)

out_array
(0 to 15)

```
four_to_sixteen :
process (in_array)
        variable sum : INTEGER RANGE 0  to 15 ;
begin
        sum := 0 ;
        sum_of_in :
                for i in in_array'RANGE loop
                        sum := sum + (2**i)*in_array(i) ;
                end loop sum_of_in ;
        zero_array :
                for j in 0 to 15 loop ;
                        out_array(j) <= 0 ;
                end loop zero_array ;
        out_array(sum) <= 1 ;
end process four_to_sixteen ;
```

10.6 Four-to-sixteen Decoder

Consider the following model specification:

Create a model of a device which has four lines of input and sixteen lines of output. The function of the device is to make the sixteen output lines high based on the pattern of the input lines. This is a basic binary decoder. The input lines represent a binary number. The output lines represent the decimal equivalent. The four-bit binary number can represent the numeric values zero to fifteen. Therefore, there needs to be sixteen output lines. The high line represents the (decimal) value concept of the binary code.

The model of Illumination Box 10.6 shows one approach for creating this functionality. To reduce the length of the model, the input array in_array is defined as an array of integers. Each element in the array can store the integer numbers 0 or 1 (this maintains the idea of bits) and there are four elements. The output array is also an array of integer numbers (limited to either 0 or 1), but contains sixteen elements.

1. The body of the model consists of a single process which is sensitive to the input array. A variable called sum is declared with the process. This variable is used to hold the calculated sum of the input array.
2. When the process is activated, sum is set to zero. This is important because the initial value of variables is set only upon invocation of the model. After invocation, the variable retains its last value unless reset to a new value. If sum were not reset each time the process activates it would continue to grow larger each time through the process.
3. The first **for** loop converts the values in the input array to a decimal value. Notice the use of the array attribute 'RANGE. This determines the size of the array and sets the number of times the loop is repeated. In this case 'RANGE returns 0 to 3. The statement sum := sum + (2 **i)*in_array(i); is used to add up the value of the input array by multiplying element times its positional value. For example when i is 0, the value in in-array (0) is multiplied by 1, the positional value of the element.
4. The second **for** loop fills the output array with zeros. This initialisation is necessary to delete the previous value of the output.
5. Finally, the output array element which represents the input code is turned on. The value of the variable sum indicates which of the output elements should be assigned the 1.

Instead of the input and output being arrays of integers, arrays of bits could have been used. The only change needed for the output array would be to place single quote marks around the 0 and 1.

10.7 ASSERT Statement

- **Checks condition and reports message with severity level if condition is not true**

- **Format**

    - **ASSERT** condition ;

    - **ASSERT** condition **REPORT** "message" ;

    - **ASSERT** condition **SEVERITY** level ;

    - **ASSERT** condition **REPORT** "message"
      SEVERITY level ;

- **Example**

```
ASSERT signal_in = '1'
        REPORT "Input is not 1"
        SEVERITY WARNING ;
```

10.7 ASSERT Statement

There are places within a model where the modeller knows what the value of certain signals and variables should be. This knowledge is used during the testing of a model to locate errors. As the model is executed, object values are displayed in the monitor window or in the transcript window (with the examine command). If the value of the object is not what is expected an error may exist in the model segment which was just executed.

This process can be long and tedious and, while it may be suitable during the testing phase, it is not necessarily suitable for error tracking during a job-related simulation. The ASSERT statement in VHDL helps to "automate" this process. This statement checks a condition and reports a message with a severity level if the condition is not true. The effect is to say to the model "At this point this should be true. If it is not true, this is what I want you to tell me and what you want me to do". This statement can appear in four ways:

1. ASSERT condition;
   An ASSERT statement cannot be written without a condition. This condition must be a Boolean expression which expresses the states which must be true at this point in the model. Typically, the condition consists of relational expressions showing the expected value of certain objects, e.g., object = '1'
   This condition says that the expected value of object is '1'. If the actual value is not '1' a message appears in a report window and a severity level is sent to the simulator. Since a message and a severity level do not appear in this form of ASSERT the default message "ASSERT VIOLATION", and the default severity level, ERROR, are used.
2. ASSERT condition REPORT "message";
   This form shows the addition of the REPORT clause. The message which appears after the key word REPORT is displayed in a report window during simulation. It takes the place of the default message. The quote marks are shown to stress that the message must appear in quote marks.
3. ASSERT condition SEVERITY level;
   This form shows the use of the SEVERITY clause. The severity level which is reported to the simulator is the value indicated by level. This value replaces the default value, error. This is very important since certain severity level values stop simulation and others allow it to continue.
4. ASSERT condition REPORT "message" SEVERITY level;
   This shows the ASSERT statement with both options. Everything said about these options in points 2 and 3 apply in this case. It is important

10.8 SEVERITY Levels

- Levels of severity from least to greatest :
    - NOTE - general information
    - WARNING - undesirable condition
    - ERROR - task completed, result wrong
    - FAILURE - task not completed
- Simulation stops when severity matches, or exceeds specified **SEVERITY** level
- Severity level which will stop simulation will be specified by a System Command such as **ASSERTION SEVERITY**
- An assertion severity may default to **ERROR** in some systems, e.g. in Mentor Graphics Quicksim

to make the distinction between "stops" simulation and "terminates" simulation. While ASSERT statements do not terminate the simulation they may stop it.

10.8 Severity Levels

The severity level sent by the ASSERT statement to the simulator is extremely important. It is used, for example, by Mentor Graphics System-1076 to determine whether simulation should be continued or be terminated. The four severity levels specified by IEEE LRM, in ascending order of severity, are:

1. NOTE
   This is the "lowest" level of severity. It indicates that the ASSERT is for general information only. In the default configuration, this level of severity does not stop the simulation.
2. WARNING
   This level is used with undesirable conditions. In the default configuration, this level of severity does not stop the simulation.
3. ERROR
   This level is used most often to indicate that the task was completed but that the result is known to be wrong. In the default configuration, this level of severity should stop the simulation.
4. FAILURE
   This level indicates that the task was not completed. In the default configuration, this level of severity should stop the simulation.

It should be possible to set which levels of severity will stop the simulation and which will not. Simulation should stop when the severity level matches, or exceeds, a specified severity level.

ASSERT *statements do not terminate simulations. They simply stop simulations.*

Simulation can usually be restarted by a System Command such as RUN, with an appropriate time period stipulated. Any problem values causing the simulation to stop can usually be reset using a command such as SET.

10.9 Placing ASSERT Statements

- **May appear within**
  - **concurrent statement areas**
  - **sequential statement areas**
  - **statement of entity declaration**

- **Example**

```
entity rs_flip_flop is
        port (r, s : in BIT ;
                        q, not_q :  out BIT) ;
end rs_flip_flop ;
architecture behaviour of rs_flip_flop is
begin
        ASSERT not (r = '1' and s = '1')
                REPORT "race condition"
                        SEVERITY FAILURE ;
                                •
                                •
                                •
```

10.9 Placing ASSERT Statements

ASSERT statements have both a concurrent and a sequential form. Therefore these statements may be placed in either statement area. In addition, they may be placed in the statement area of the entity declaration. Since this area has not been, so far, discussed the use of the ASSERT statement in the entity declaration area will be postponed until later.

1. Concurrent assertion

   Concurrent ASSERT statements have an associated sensitivity list. These statements are sensitive to the signal objects appearing in their condition. Consider the statement:

   ```
   ASSERT a = '1' ;
   ```

   This statement is sensitive to signal *a*. Whenever there is an event on signal *a* this statement is evaluated. The effect would be to issue messages and severity levels whenever signal a changed to something other than '1'.

2. Sequential assertion

   Sequential ASSERT statements are evaluated whenever they are encountered. Since these statements do not have a sensitivity list associated with them they execute only when the code sequence in which they are contained is executed. Consider the statements:

   ```
   example: process (b)
            begin
                    ASSERT a = '1' ;
            end process;
   ```

   Even though the ASSERT statement in this example checks to see if the value in object *a* is '1', the assertion is not checked unless there is a change in signal *b*.

There are many occasions when the condition in an assertion statement may need to check the value of several objects. This is done by combining relational and logical operators. For example: ASSERT **not** (r = '1' **and** s = '1') REPORT "race condition" SEVERITY failure;

If both inputs to an R-S flip-flop go to '1' the output is indeterminate i.e. is in an unstable state. A model of this device should warn if a race condition exists and should stop the simulation. The above assertion statement accomplishes both of these tasks. The condition checks to see if the *r* value is '1' and the *s* value is '1'. Both values must be '1' for the **and** to produce a TRUE. Since the ASSERT is activated when the result is FALSE, the **not** operator is used to invert the result. Hence the message "race condition" appears in the transcript when both values are '1' and the simulation is stopped (severity level of failure).

10.9.1 ASSERTed Decoder

BIN/DEC

in_array (0 to 3)

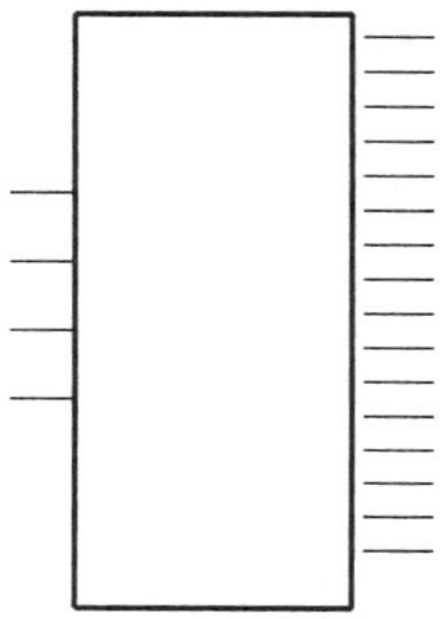

out_array (0 to 15)

```
four_to_sixteen decoder ;
 process (in_array)
        variable sum : INTEGER RANGE 0 TO 15 ;
begin
        ASSERT not    (in_array(0) = 'X' or
                       in_array(1) = 'X' or
                       in_array(2) = 'X' or
                       in_array(3) = 'X' )
                            REPORT "unknown value"
                            SEVERITY ERROR ;
        sum := 0 ;
 for  i in in_array'RANGE
                   •
                   •
```

10.9.1 ASSERTed Decoder

The model shown in Illumination Box 10.9.1 shows how an ASSERT statement might be used in the four_to_sixteen decoder model which appeared in Illumination Box 10.6. The input to the model was an array of four values. These values represented a four bit binary number. The purpose of the model was to use the value represented by the four bit binary code to turn on (drive to '1') one of the sixteen output lines. The output lines represented the numbers zero through fifteen.

To reduce the complexity of the model the elements in the input array were assumed to hold integer numbers. These numbers were limited to the values 0 and 1 to emulate binary digits. However, this may not be possible in the particular simulator being used. For example, in a Mentor Graphics simulator input values must be in terms of qsim_states. This being the case the elements may contain 'X' in addition to '0' and '1'.

'X' values cannot be interpreted in the conversion process, but one way to deal with this problem would be to place 'X' in all of the output elements. Another way would be to "trap" the 'X' values before entering the process. Consider the statement:

```
ASSERT not (  in_array (0) = 'X' or
              in_array (1) = 'X' or
              in_array (2) = 'X' or
              in_array (3) = 'X' or )
                        REPORT "unknown value"
                        SEVERITY ERROR;
```

This statement shows how an ASSERT statement could be used to check each element in the array in_array. If any of the element values are 'X' the message "unknown value" will appear in the transcript window and the simulation will be halted. A similar statement could be used to çheck for 'Z', or the condition in this statement could be expanded to include comparisons for 'Z'.

The statement has what appears to be a "reversed" notation. The **not** in front of the condition reverses the outcome of the comparison. This is necessary because of the way the check is set up. Remember that an assertion statement reports when the condition is FALSE. The checks in this statement produce a TRUE if any of the element values are 'X'. Therefore, it is necessary to change the TRUE to a FALSE to cause the report to be made and the severity to be set.

10.10 Exercises

10.10.1 Create a model of an eight bit serial-to-parallel shift register with clear. Input is serial, output is parallel. The process is keyed to the rising edge of a clock(defined as an '0' to '1' transition). If the 'clear' signal is '1', at the rising edge all outputs are set to '0'. If the 'clear' is '0', the output is shifted and new input assigned to output.

10.10.2 Create a model of an or_gate which uses assert statements to check for spikes on input signals. For this model, assume that a spike is any signal change with a duration of 5 ns or less. Test the model with the severity level of the assert statement set to NOTE, the change the severity level of the assert statement to ERROR and re-test the model.

10.10.3 Create a model for an 8-bit shift/storage register with 3-state outputs (74299 equivalent).
The pin names are to be as follows:

CP	Clock Pulse (active positive-going edge) Input
DS0	Serial Data Input for Right shift
DS7	Serial Data Input for Left Shift
I/On	Parallel Data Input or Parallel Output (3-State)
OE1_bar, OE2_bar	3-State Output Enable (active LOW) Inputs
Q0, Q7	Serial Outputs
MR_bar	Asynchronous Master Reset (active LOW) Input
S0, S1	Mode Select Inputs

FUNCTION TABLE

INPUTS								RESPONSE
MR_bar	S1	S0	OE1_bar	OE2_bar	CP	DS0	DS7	
L	X	X	H	X	X	X	X	Asynchronous reset
L	X	X	X	H	X	X	X	Q0 = Q7 = LOW
L	H	H	X	X	X	X	X	I/O Voltage undetermined
L	L	X	L	L	X	X	X	Asynchronous reset
L	X	L	L	L	X	X	X	Q0 = Q7 = LOW I/O Voltage LOW
H	L	H	H	X	↑	D	X	Shift Right D → Q0; Q0 → Q1; etc.
H	L	H	L	L	↑	D	X	Shift Right D→Q0 & I/O0; Q0→ Q1& I/O1 etc;
H	H	L	X	H	↑	X	D	Shift Left D→ Q7; Q7 → Q6; etc.
H	H	L	L	L	↑	X	D	Shift Left D→Q7 & I/O7; Q7→ Q6 & I/O6 etc.
H	H	H	X	X	↑	X	X	Parallel Load; I/On→ On
H	L	L	H	X	X	X	X	Hold: I/O Voltage
H	L	L	X	H	X	X	X	undetermined
H	L	L	L	L	X	X	X	Hold: I/On = Qn

10.10.4 Add ASSERT statements to Ex. 10.10.3 in order to observe activity and define constraints given the following additional ac characteristics for the shift register: Clock to Q0 or Q7; tphl = 26 ns, tplh = 22 ns. Clear to Q0 or Q7; tphl = 27 ns. Clock to I/O0 - I/O7; tphl = 26 ns, tplh = 17 ns. Clear to I/O0 - I/O7; tphl = 26 ns. Output enable time; tpzh = 13 ns, tpzl = 19 ns. Output disable time; tphz = 10 ns, tplz = 10 ns.

Chapter 11

Subprograms

Overview

This chapter introduces and discusses the concept of subprograms. Subprograms are sequences of statements which can be referenced from most locations within the model. They provide a method for reducing the duplication of statements in the model.

In this chapter you will learn

- That there are two types of subprogram in VHDL, functions and procedures
- That functions produce a single return value and are invoked by an expression
- That procedures can affect many values and are invoked by a statement

11.1 Subprograms

- Sequences of statements which can be executed from multiple locations in model
- Provide a method for breaking large segments of code into smaller segments
- Types of subprogram
  - Functions
    Used to calculate and return one value
  - Procedures
    Used to define an algorithm which affects many values or no values

11.1 Subprograms

Subprograms are sequences of statements which can be referenced from most locations within the model. They provide a method for reducing the duplication of statements in the model.

The modeller may require that identical statement sequences appear in several places within a model. When this happens, subprograms provide a way in which the multiple statement sets can be replaced by one set and multiple references, or calls, to that set. This reduces the apparent length of the model and makes it easier to read and interrupt. The effect is that large segments of code are reduced in size and complexity.

Subprograms are collections of sequential statements which can be referenced by an assigned identifier. Concurrent statements cannot appear within a subprogram.

However, reference to subprograms can be made in both concurrent areas and sequential areas of the model.

There are two categories of subprograms in VHDL:

1. Functions
   Functions can calculate and return a single value. They cannot affect the parameters passed to them.
2. Procedures
   Procedures may change parameters passed to them. Therefore they have the ability to affect many values in the calling statement.

11.2 Functions

- Provide a method of performing complex algorithms
- Produce a single return value
- Invoked by an expression
- Must contain a return clause
- May consist of two parts
  - Declaration
  - Body
- May not affect parameters

11.2 Functions

A subprogram is a sequence of statements which performs a particular task. In some ways it may be thought of as a model within a model. The task it performs can be a very simple, or a very complex, algorithm.

The modeller often requires that identical blocks of functionality appear, in several places, within a model. When this happens, functions provide a way in which the multiple statement sets can be replaced by an expression which references, or calls to, that set. This reduces the apparent length of the model and makes it easier to read and to interrupt.

Functions are one way of structuring a subprogram. When a function is referenced (called), the values of certain objects in the model are passed to the function. These objects are called the passing parameters. A feature of functions is that these values may not be changed in the function. Instead, the function consists of sequential statements which use the values of the passing parameters to calculate a new value. This new value is returned to the point in the model where the function was called.

Functions have several unique features; these are:

1. They are invoked by an expression - as opposed to a statement.
   Consider the statement: signal_a <= function_a(signal_b, signal_c) ;
   This function is called by its name, function_a. This use of the name of the function with associated object names (*signal_b, signal_c*) within a statement is referred to as an expression.
2. They produce a single return value.
   When a function is executed one, and only one, value is returned to the calling statement. In the example above one value would be returned by the function call. This value would be assigned to *signal_a*.
3. They must contain a **return** clause.
   Since functions return a single value there must be some method of indicating that the process has been completed and what value is to be returned. The **return** clause serves this purpose.
4. They may not affect the values of the objects which are passed to them.
   The value of the objects (parameters) in the calling expression (for example, *signal_b, signal_c*) cannot be changed.
5. They must have a "body" and may have a "declaration".

11.2.1 function Declaration

- **Specifies**
  - **Function name**
  - **Input**
  - **Type of return value**
- **Must**
  - **Exist for recursive functions**
  - **Exist when body occurs below first function call**
  - **Occur above first function call**
  - **Appear in same statement declaration area as function body**
- **Format**

  function name (parameter_list) return type;

- **Example**

  function exam (**signal** signal_a : BIT)
  return BIT ;

11.2.1 function Declaration

Functions may consist of two parts, a function declaration and a function body. A function body is required if the function is to have any purpose. The function declaration may, or may not, be required.

A function declaration has the following format:

function name (parameter_list) **return** type ;

and provides three pieces of information.

1. The function name.
   The function name is the name used when calling (invoking) the function. For example, **function** exam identifies a function named exam.
2. The parameter list identifies the objects which are used in the function to hold the values passed to the function from the calling point. This list states the class, name, mode and type of the object. For example: (**signal** signal_a : **in** BIT) identifies a signal object which cab hold values of type BIT.
3. The **return** clause identifies the type of the return value.
   The type identifies the set of values which can be returned by the function. For example: **return** BIT indicates that this function can return either a '0' or a '1'.
   As was pointed out previously, the function declaration does not have to exist. Usable functions may be written without a function declaration. However, a function declaration must exist when:
   i) The function is recursive.
   If a reference to the function is made within the body of the function (the function calls itself) a function declaration must exist.
   ii) The body of the function occurs below the first call to the function.

Functions have to be declared. They are declared in the declarative areas. However, since functions can be called from other functions or procedures declared in the same declarative area, there must be some method of indicating that, although the function body has not yet been declared, it will be declared. The function declaration serves this purpose.

Two very important points should be remembered. When a function declaration is made, it must be made in the same declarative area as the function body (NB. This is not strictly true in the case of packages where the function declaration appears in the package declaration and the function body appears in the package body). Function declarations always precede the first call to the function.

11.2.2 Parameter List

- **Identifies class, name, mode and type of objects which contain values passed to function**

- **Format**

  (CLASS object_name : MODE **type**)

- **Object class may be signal or constant**

- **Type of objects must correspond to type of values passed**

- **Examples**

  function exmpl (**signal** signal_a : **in** BIT) . . .
  function exmpl (**signal** signal_a : BIT) . . .
  function example (signal_a : BIT) . . .*

*** CLASS of object defaults to constant if not stated**

11.2.2 Parameter List

The parameter list which appears in the function declaration was briefly introduced in the previous discussion. It identifies the class, name, mode and the type of objects which contain values passed to the function. The format of the parameter list is:(CLASS object_name : MODE **type**)
The parameter list provides these pieces of information.

1. CLASS
   The object class must be either **signal** or **constant**.

   Variables are not allowed in functions.

   If the class is omitted from the parameter list the class of object is assume to be **constant** (i.e. the default is **constant**).
2. object_name
   The object name is supplied by the modeller. It can be any name which follows the standard VHDL naming rules, This name does not have to match any other object name in the model.
3. MODE
   The mode identifies the direction of data flow with regard to the function.

   in *is the only mode allowed in functions.*

   Therefore, the mode is usually omitted from the parameter list and becomes **in** by default. Because the mode is **in** none of the objects in the parameter list can be modified (NB. Only the values **in** or **inout** can appear to the right of an assignment operator).
4. **type**
   The type identifies the value set which the object can contain. Values which do not appear in the value set cannot be passed from the model to the function. Therefore, the type of the object in the parameter list must match the type of the object, from which the value is being passed, in the model.

The effect of the parameter list is to declare the objects which contain the values passed to the function. Because the only mode allowed is **in**, the mode is usually omitted from the list. The class could also be allowed to default to **constant**. It may be tempting to do this since object values cannot be modified in the function (cf.. Section 5.1 Objects). However, there are valid reasons for declaring an object to be of the class **signal**. The primary reason is that signals have associated attributes which may be of use in the function.

11.2.3 return Type

- **Identifies type of value returned by function**
- **Type must be declared**
- **Return value must be of same type as identified by "return** type"
- **Format**

  . . . **return** type
- **Examples**

  . . . **return** BOOLEAN
  . . . **return** BIT

11.2.3 return Type

The return type identifies the value set from which the value returned by the function can be selected. Its format is:

return type

This clause does not identify the specific value that is to be returned. Instead, it states the value set from which a value may be selected for return. This means that the return value must be one of the values identified by the type.

A type must be identified for the return value. There is no default type. When deciding on the return type it is important to consider the use of the return value. The type must match the use. For example, a type BIT could not be returned to a calling statement which assigned the return value to a signal of type BOOLEAN.

11.2.4 function Body

- **Defines how function behaves**
- **Consists of sequential statements**
- **Placed in declarative statement areas**
- **Returns only one value**
- **Must contain a return statement**
- **Format**

```
function name (parameter_list)
                    return type is
      -- declarative_statements
begin
      -- sequential statements
      -- including return expression
end name ;
```

11.2.4 function Body

The body of the function defines how the function behaves. If a function is to have a purpose it must have a body.

A function body consists of sequential statements describing the relationship between the input parameters and the return value.

Concurrent statements cannot appear in a function.

Any of the sequential statements which have been discussed in this book may appear in a function. There must be at least one **return** statement in the body. This statement identifies the value to be returned to the model when it is executed. Although there may be multiple return statements in the function only one value is returned. **return** not only identifies the value to be returned, it also terminates the function.

The function body is placed in the declarative statement areas. The body can exist and be accessed without the existence of a function declaration. However, if a function declaration does not exist, the body must be placed at the same level, or at a level above, the area where it is called. This means that a function body placed in one process cannot be accessed from another process (unless a function declaration for the body is placed in the declarative area of the architecture).

The function body format is:

```
function name (parameter_list) return type is
        -- declarative_statements
begin
        -- sequential_statements
        -- including return expression
end name ;
```

Note the similarity between the first line of the function body and the function declaration. Except for the **is** at the end of the line of the body statement the two are identical. It is very important that the name, parameter list and return type match if both a body and a declaration appear in the model. The purpose of the declaration and the body line are the same.

Because the objects which appear in the parameter list cannot be modified, and because there may be a need for objects which can be modified in the function, there is a declarative area in the function. Objects declared in this area can only be used in the function. They may not be transferred to, or used outside, the function.

The behaviour of the function is described between the **begin** and **end** name; of the function. These statements must be sequential statements and at least one **return** statement must appear somewhere within them. The **return** statement

11.2.5 return Statement

- **returns value to object assigned to receive function output**
- **Terminates subprogram execution**
- **Used only in subprograms**
- **Format**

```
functions
    return expression ;

procedures
    return ;
```

- **Example**

```
function example (a, b : in BIT)
    return BOOLEAN is
begin
    return a = b ;
end example ;
```

is discussed in detail in Illumination Box 11.2.5. It should be noted that the **return** statement is not the return clause which appears in the first line of the body. One identifies the value set which the return value can be drawn from; the other identifies the specific value which is to be returned.

It must be stressed that the "name" of the function <u>must</u> appear after **end**. This is not an optional usage such as it is after **end** in an entity and its architecture.

11.2.5 return Statement

The **return** statement has two purposes:

1. To return the value created by the function to the object assigned to receive the output of the function.
2. To terminate the execution of a subprogram.

This VHDL statement appears only in subprograms. Because there are two types of subprogram there are two formats of the **return** statement:

1. Functions.

   return expression ;

   This format includes an expression which identifies the value to be returned by the function. The expression can be a value, an object identifier or an expression which evaluates to a value. Its purpose is twofold: first, to return a value to the calling area and second, to terminate the execution of the function.
2. Procedures.

   return ;

   This form does not include a return expression and is, therefore, used exclusively with procedures (procedures do not return a specific value: rather they can affect none, some or all of the passing values). The purpose of **return** is to terminate the execution of the procedure.

The following example shows how **return** might be used in a function:

```
function example (a, b : in BIT) return BOOLEAN is
begin
    return a = b ;
end example ;
```

This entire function consists only of a **return** statement. The values in the objects *a* and *b* are compared using the equals (=) relational operator. The result will be either TRUE (they are equal) or FALSE (they are not equal). The TRUE or FALSE will be returned to the place from which the function was called. Since the return value set is identified as boolean values in the **return** clause, this is a valid expression.

11.2.6 transfer_check Function

```
function transfer_check
    (
    signal a_ray : in BIT_VECTOR (0 to 7) ;
    signal b_ray : in BIT_VECTOR (0 to 7)
    )
    return BOOLEAN is
     variable sum1, sum2 : INTEGER := 0 ;
begin
        for i in a_ray'RANGE loop
            if a_ray (i) = '1' then
                sum1 := sum1 + 1 ;
            end if ;
            if b_ray (i) = '1' then
                sum2 := sum2 + 1 ;
            end if ;
        end loop ;
        return sum1 = sum2 ;
end transfer_check ;
```

Multiple **return** statements can appear in a function. This is a useful technique when different values are to be returned in different instances. However, only one **return** statement will be executed. This is because the **return** statement which is evaluated first terminates the execution of the function.

11.2.6 transfer_check Function

The model shown in Illumination Box 11.2.6 illustrates how a simple function might be written. This function performs a transfer check. The values of two arrays are transferred into objects in the function. Array *a_ray* represents an initial array. Array *b_ray* represents an array which has had the values of *a_ray* read into it. This function checks to see that the number of '1's is the same in both arrays. The assumption is that if the transfer values between two bit arrays is done correctly the number of '1's should be the same.

The check is made by counting the number of '1's in each array. Since values in the parameter list cannot be changed, and since a counting process requires the adding of values to an object, this process requires the declaration of two variables. Variable sum1 is used to hold the number of '1's in *a_ray*. Sum2 is used to hold the number of '1's in *b_ray*. The actual summation is done in a **for** loop. Since both arrays must be the same length the loop is performed 'RANGE times. This ensures that all elements of the array are examined. If a '1' is encountered in *a_ray*, 1 is added to sum1; if a '1' is encountered in *b_ray*, 1 is added to sum2. When the loop is exited from the boolean result of sum1 = sum2 is returned to the calling point in the model.

An interesting point about functions can be seen in this transfer_check model. Variables in functions are re-initialised each time the function is called. Consider the following:

```
function totaler (
                a_ray : in BIT_VECTOR (0 to 10)
                )
    return  integer is
    variable sum : INTEGER := 0 ;
begin
    for i in 0 to 10 loop
            if a_ray (i) = '1' then
                    sum := sum + 1 ;
            end if ;
    end loop ;
    return  sum ;
end totaler ;
```

11.2.7 function Calls

- **Stated as expressions**
- **Case execution of function body**
- **Specify passing parameters**
- **Occur at or below level of function declaration**
- **Format**

  function_name (passing_parameter_list)

- **Example**

  if transfer_check (array_1, array_2) **then**. . .

```
summer : process (a_ray)
        variable sum : INTEGER := 0 ;
    begin
        for i in 0 to 10 loop
            if a_ray (i) = '1' then
                sum := sum + 1 ;
            end if ;
        end loop ;
    end process summer ;
```

The process example must have the variable sum set to zero each time it executes. If sum was not reset its initial value would be the value it had at the end of the last execution of the process. Thus sum would not represent the number of '1's in the array.

11.2.7 function Calls

A function is invoked by a calling expression. This expression is termed a "function call" and causes execution of the function body.

The function call for a particular function can occur at the same level as the function declaration, or at any level below the declaration level. For example, if the declaration for a function occurs in the declarative area of the architecture the function may be called from any concurrent, or sequential, statement appearing in the architecture. However, if the function occurs in a declarative area within a process in the architecture it can only be called from within that process; the function cannot be called from another process in the architecture.

The same rules apply when only the function body exists. Calls to the function can be made at the same level, or lower levels. The advantage in using both the body and the declaration is that the declaration can exist at a higher level than the body to allow use of the body across parallel structures. For example, by using a declaration, the function can be called in other function bodies which occur after the declaration but before the body.

Function calls are made by stating the name of the function and a list of the objects whose values are to be passed to the function. The format is:

```
function_name (passing_parameter_list)
```

Since functions return a single value, a function call can be made in any instance where it could be replaced by a single value of the type returned by the function. For example, if the return type of the function was a boolean value, the function call could be made as the condition in an **if** statement.

For example:

```
if transfer_check (array_1, array_2) then. . .
```

The name transfer_check refers to the example shown in the previous discussion.

11.2.8 Passing Parameter List

- **Specifies objects from which values pass to function**

- **Positional association**
  - **Parameters in call match order of parameters in function parameter list**
  - **Example**

```
function tran_ck (signal a_1 : in BIT ;
                  signal b_2 : in BIT)
                  return BOOLEAN is . . .
        •
        •
if tran_ck (ray_1, ray_2) . . .
```

- **Named association**
  - **Relates parameters in call by name to parameters in function list**
  - **Example**

```
function tran_ck (signal a_1 : in BIT ;
                  signal b_2 : in BIT)
                  return BOOLEAN is . . .
        •
        •
if tran_ck (b_2 => ray_2, a_1 => ray_1) . . .
```

Since this function returns a boolean value, this is a valid use of the function call. The values passed to the function come from the two objects array_1 and array_2. These values are passed to the objects a_ray and b_ray in the function. The function compares the number of '1's in each of these objects and returns a TRUE or FALSE based upon whether or not the numbers of '1's are equal.

11.2.8 Passing Parameter List

The parameter list in the function call expression is called the passing parameter list, or formal parameter list. The passing parameter list specifies the objects form which values pass to the function. There are two ways in which objects in this list can be associated with objects in the function parameter list. They are:

1. Positional Association

   In positional association the objects named in the function call expression are matched by position to the objects in the function parameter list. For example, consider the function whose first line is

   function tran_ck (**signal** a_1 : **in** BIT ;
   signal b_2 : **in** BIT) **return**
   BOOLEAN **is** . . .

   and the function call :

   if tran_ck (ray_1, ray_2) . . .

   The value *ray_1* would be associated with *a_1* in the function and the value in *ray_2* would be associated with *b_2*. The association is done by position. *ray_1* is the first object in the formal parameter list of the calling expression and *a_1* is the first object in the function parameter list. *ray_2* and *b_2* are the second objects respectively.

2. Named Association

   The named association matches the objects in the function call expression to the objects in the function parameter list by name. For example, a named call to the above function example would be :

   if tran_ck (b_2 => ray_2, a_1 => ray_1) . . .

   In this example it is easy to see that *b_2* is associated with *ray_2* and that *a_1* is associated with *ray_1*.

   NB. The symbol (=>) is an association symbol, and while it may appear that *b_2* is being transferred to *ray_2*, in fact, the reverse is true; the symbol is an 'arrow' symbol showing that the value from *ray_2* is being transferred to *b_2*. Also note that the order of the object associations has no importance in named association.

It is possible to mix positional association and named association in a function call expression. However, all positional associations must come first and once named association is used all remaining associations must be made by name.

11.3 bit_to_boolean

```
entity example is
    port (in_a : in BIT ;
          out_b : out BOOLEAN) ;
end example ;

architecture func of example is
    function bit_to_boolean
          (bit_in : in BIT)
                return BOOLEAN is
    begin
          if bit_in = '1' then
                            return TRUE ;
          else
                            return FALSE ;
          end if ;
    end bit_to_boolean ;

    begin
          out_b <= bit_to_boolean (in_a) ;
    end func ;
```

It is never possible to have more objects in the formal parameter list than in the function parameter list.

11.3 bit_to_boolean

The model in Illumination Box 11.3 has one input line and one output line. The input line, in_a, carries signals of type BIT. The output line out_b carries signals of type BOOLEAN. The purpose of the entity is to convert signals of type BIT to signals of type BOOLEAN. In this transform '1' equals TRUE and '0' equals FALSE.

There are several things to be pointed out concerning this model:

1. Function declarations and function bodies are placed in declarative regions.
   Notice that this function is placed in the declarative region of the architecture.
2. Functions may be written without a function declaration if the call to the function exists at or below the body.
   This function has only a body. Notice the placement of the body. Because the body occurs in the declarative region of the architecture, the function may be called from any structure which occurs within the architecture.
3. Functions are called by placing the function name before a parenthetical list of the object values to be passed to the function.
   There is only one line in the activity area of this model. It consists of a signal assignment statement which assigns the return value from the function to the output port. The function is called by name and the value of the input port is passed to the function.
4. Functions may have more than one **return** statement but only one **return** statement will be executed.
   This function has two **return** statements. The first returns the boolean value TRUE and the second returns the value FALSE. Although these **return** statements appear in an **if** statement that ensures only one of them is executed, when a **return** statement is executed control is transferred from the function back to the calling statement.

Each of these facts has been presented, and discussed previously. This model is designed to show, and reinforce, the practical understanding which has been developed in theoretical discussion.

11.4 Procedures

- Provide a method of performing complex algorithms
- May produce multiple output values
- May affect input parameters
- Are invoked by a statement
- May consist of two parts
  - Declaration
  - Body
- May contain a return statement

11.4 Procedures

Procedures are similar to functions in some aspects and very different in others. Both are subprograms which provide a method of performing complex algorithms. Both may have declarations and bodies. However, the differences are more striking than the similarities, viz. :

Functions cannot change the values of parameters passed to them; procedures may change the values of the parameters passed to them.

This is the fundamental difference between functions and procedures.

Functions must contain a return statement; procedures may contain a return statement.

This is a symptom of the fundamental difference between functions and procedures. Since functions cannot change the values of parameters passed to them, the only way in which they can demonstrate activity is to return a created value. Procedures can change the values of the parameters passed to them and, therefore, do not need to return a specific value in order to demonstrate their activity. In point of fact procedures *cannot* return a specific value; activities in procedures are demonstrated by *changes* in the values of passing parameters. If a return statement appears in a procedure it is used to terminate the procedure and not to return a specific value.

Functions return a single value; procedures may produce multiple output values.

Procedures can produce many values which are returned via the passing parameter list. Objects which appear in the procedure parameter list with a mode of **out** are passed from the procedure to the model. Objects of this mode do not receive values from the model when the procedure is called; therefore, they are effectively created by the procedure.

Functions cannot affect the values of objects passed to them; procedures may affect the values of objects passed to them.

All object values passed to functions have a mode of **in**. Values of objects of mode **in** cannot appear to the left of assignment operators and, therefore, may not be changed. Object values passed to procedures may have modes of **in, out** or **inout**. While those values of mode **in** may not be changed, the object values of mode **inout** may be modified and returned to the calling statement.

Functions are invoked by expressions; procedures are invoked by statements.

This difference means that a function call may be incorporated into a more complex statement such as a signal assignment statement, but the procedure call must exist as its own statement.

11.4.1 procedure Declaration

- **Specifies**
  - **Procedure name**
  - **Parameters**
- **Must**
  - **exist for recursive procedures**
  - **exist when body occurs below first procedure call**
  - **appear in the same statement declaration area as procedure body**
- **Format**

```
procedure name (parameter_list) ;
```

- **Example**

```
procedure example (variable a : in BIT ;
                   signal b : out BIT) ;
```

11.4.1 procedure Declaration

Procedures, like functions, may consist of two parts, a procedure declaration and a procedure body. A procedure body is required if the procedure is to have any purpose. A procedure declaration may, or may not, be required.

A procedure declaration has the following format:

procedure name (parameter_list) ;

and provides these two pieces of information:

1. The procedure name.
   The procedure name is the name used when invoking the procedure. For example:

   procedure example (**signal** a : **in** BIT) ;

   identifies a procedure named example.
2. The parameter list.
   The parameter list identifies the objects which are used in the procedure to hold the values passed to the procedure from the main model. This list states the class, the name, the mode and the type of the object. For example:

   (**signal** a : **in** BIT)

   identifies a signal object which can hold values of type BIT.

The procedure declaration does not have to exist, valid procedures may be written without a procedure declaration. However, a procedure declaration must exist when:

a. The procedure is recursive.
   If a reference to the procedure is made within the body of the procedure (i.e. the procedure refers to itself) a declaration must exist.
b. The body of the procedure occurs below the first call to the procedure.
   Procedures are declared in declarative areas. However, since procedures can be called from other procedures or functions declared in the same declarative area, there must be some method of indicating that, although the body has not yet been declared, it will be. The procedure declaration serves this purpose.

Three very important points are made in these last two items. First, when a procedure declaration is made, it must be made in the same declarative area as the procedure body (this is not strictly true in the case of packages where the procedure declaration appears in the package declaration and the procedure body appears in the package body). Second, procedure declarations are not necessary unless the procedure call precedes the body of the called procedure. Third, procedure declarations, if used, always precede the first call to the procedure.

11.4.2 Parameter List

- **Identifies class, name, mode and type of objects which contain values passed to and from procedure**

- **Mode determines what actions can be taken on parameter**

- **Object class may be**
  constant, signal or variable

- **Mode may be**
  in, out or inout

- **Format**
  . . .(CLASS object_name : MODE **type**) . . .

- **Examples**
  procedure a (**variable** a : **in** REAL) . . .
  procedure b (**signal** signal_a : BIT) . . .
  procedure c (signal_a : BIT) . . .

11.4.2 Parameter List

The parameter list which appears in the procedure declaration and procedure body is similar to the parameter list which appears in functions. It identifies the class, name, mode and type of objects which contain values passed to the procedure. The effect of the parameter list is to declare the objects which contain values passed to the procedure.

The format of the parameter list is:

(CLASS object_name : MODE **type**)

The parameter list provides these pieces of information:

1. CLASS

   The object class may be **signal, variable** or **constant**. Notice that all three classes are allowed in procedures.

   If both the class and the mode are omitted from the parameter list the class of objects is assumed to be **constant** and the mode **in**.

   If the class is omitted and the mode is **out** or **inout** the class is assumed to be **variable**.

   There is no instance in which the class will default to **signal**.

2. object_name

   The object name is supplied by the modeller. It can be any name which follows the standard VHDL naming rules. The name does not have to match any other object name in the model.

3. MODE

   The mode identifies the direction of data flow with regard to the procedure. Three modes are allowable; **in**, **out** and **inout**. The mode determines how values are passed to the procedure and what actions can be taken on the parameter values. If the mode is omitted a value of **in** is assumed.

4. **type**

   The type identifies the value set which the object can contain. Values which do not appear in the value set cannot be passed from the model to the procedure. Therefore, the type of the object in the parameter list must match the type of the object from which the value is being passed in the model.

There is a very important point associated with the CLASS which must be stated. If the object in the parameter list is **variable** the procedure can only be used (called) from a sequential area of the model. This is because only objects of class **variable** may be passed to a variable object in the parameter list of a procedure.

11.4.3 Parameter Modes

- **Determine how parameters are accessed with procedures**

- **Modes**
  - **in** — Parameter may be read into procedure but may not be modified
  - **out** — Parameter is created in procedure for use by calling code and may not be passed down to procedure from calling code
  - **inout** — Parameter may be read from calling code, modified and returned to calling code

- **Defaults for class when mode is given**

  - class defaults to constant for mode in
  - class defaults to variable for out or inout

11.4.3 Parameter Modes

The mode which appears in the parameter list determines how parameters are accessed in procedures. It determines how values are passed to the procedure and what actions can be taken on the parameter values.

Three VHDL modes can be used in the parameter list:

1. **in**

   Parameter values are transferred into the procedure, but the value may not be modified.

   Values associated with objects of mode **in** and of type **constant** or **variable** are passed by "copy" but cannot be changed.

   Values associated with mode **in** and of type **signal** are associated with the corresponding parameter in the model and cannot be changed.

2. **out**

   Parameter value may not be transferred to the procedure from the calling code. Values for objects of this type are created within the procedure and are transferred out to the objects of the model.

3. **inout**

   Values are read from the calling code, may be modified within the procedure and are returned to the calling code.

   Values associated with objects of mode **inout** are passed by "copy" in the case of **variable**s and by association in the case of **signal**s. These values may be changed.

While it is possible to transfer a **constant** value from the model to an object of class **constant** and mode **inout** in the procedure, it is not necessary; the constant value could not be changed without violating the nature of the class **constant**.

11.4.4 procedure Body

- **Defines how procedure behaves**
- **Consists of sequential statements**
- **Placed in declarative area**
- **May read, create or modify parameters**
- **Must exist at or above the level of use if procedure declaration is not used in the model**
- **Format**

```
procedure name (parameter_list) is
        -- declarative_statements
begin
        -- sequential_statements
end name ;
```

11.4.4 procedure Body

The body of the procedure defines how the procedure behaves. If a procedure is to have a purpose it must have a body.

A procedure body consists of sequential statements describing the functionality of the procedure.

Concurrent statements cannot appear in a procedure.

Any of the sequential statements which have been discussed in this text may appear in a procedure.

return statements may appear in a procedure body, but they are not a requirement. If they do appear in a procedure body they must appear without a return value (i.e. with **return** but not with **return** value). The sole purpose of using the **return** statement in a procedure is to terminate its execution. If a **return** statement does not appear in the procedure, the procedure is terminated when the **end** is reached.

The procedure body is placed in declarative statement areas. The body can exist, and be accessed, without the existence of a procedure declaration. However, if a procedure declaration does not exist, the body must be placed at the same level or at a level above the area from which it is called. This means that a procedure body placed in one process cannot be accessed from another process - unless a procedure declaration for the body is placed in the declarative area of the architecture.

The procedure body format is:

```
procedure name (parameter_list) is
        -- declarative statements
begin
        -- sequential_statements
end name;
```

Notice the similarity between the first line of the procedure body and the procedure declaration; except for the **is** at the end of the first line of the body the two are identical. It is very important that the name, parameter list and return type be the same in both the body and the declaration. Depending upon the mode of the objects appearing in the parameter list, values passed to the procedure may or may not be modified; and can even have their source from within the procedure. Therefore, it is extremely important that the mode of the objects named in this list be carefully considered before the list is created.

The declarative area in the procedure may be used to declare objects which are used only within the procedure. Objects which are declared in this area can only be used in the procedure and may not be transferred or used outside the procedure.

11.4.5 find_minimum procedure

```
procedure find_min
    (variable a_ray : in int_array ;
     variable  min_val : inout INTEGER ;
     variable old_min : out INTEGER ) is
begin
    old_min := min_val ;
    for i in a_ray'RANGE loop
        if a_ray (i) < min_val then
         min_val := a_ray (i) ;
        end if ;
    end loop ;
end find_min ;
```

The behaviour of the procedure is described between the **begin** and the **end** name; of the procedure. The statements appearing within this region must be sequential statements. Notice that the name of the procedure must follow the **end**.

11.4.5 find_minimum procedure

The model in Illumination Box 11.4.5 shows how a simple procedure might be written. This procedure finds the minimum value stored in an array. To simplify the discussion, the size of the array is determined by the type int_array and the values stored in the array are integers. The actual number of elements in the array is not important to the discussion. Because it is important that the values stored in the input array be not inadvertently changed the input array is given a mode of **in**.

Two other objects appear in the parameter list; *min_val* and *old_min*. The signal *min_val* has the mode **inout**; this indicates that it is transferred to the procedure from the model, can be modified within the procedure and is transferred back to the model. The purpose of this objects is to hold the minimum value of all arrays which have been examined by this procedure. The second signal *old_min* has the mode **out**; this indicates that it is not transferred from the model to the procedure - it has its source in the procedure - it contains the previous, or last, minimum value.

The minimum value is determined by comparing each element in the array with the current minimum value. If a smaller value is found it is transferred to the object *min_val*. Notice that the **for** loop is done 'RANGE times; this ensures that each value in the array is examined. Also notice that no **return** statement appears in this procedure. Execution is transferred back to the model when the **end** statement in the body is encountered.

11.4.6 procedure Calls

- **Cause execution of procedures**
- **Can be used in sequential or concurrent statement areas**
- **Specify objects whose values are passed to parameters in procedure**
- **Concurrent calls are activated by changes in any signal associated with a parameter of mode in or inout**
- **Format**
  procedure_name (passing_parameter_list) ;
- **Example**
  find_min (array_1, out_1, old_1) ;

11.4.6 procedure Calls

A procedure is invoked by a calling statement. This statement is termed a "procedure call" and causes execution of the procedure body. There are both concurrent and sequential forms of the procedure calling statement.

The procedure call for a particular procedure can occur at the same level as the procedure declaration or at any level below the declaration level. For example, if the declaration for a procedure occurs in the declarative area of the architecture, the procedure may be called from any area in the architecture body and below the declaration. However, if the procedure occurs in a declarative area within a process in the architecture the procedure cannot be called from outside that process (for example, from another process in the architecture).

The same rules apply when only the procedure body exists. Calls to the procedure can be made at the same or lower levels. The advantage to using both the body and the declaration is that the declaration can exist at a higher level than the body to allow use of the body across parallel structures.

Procedure calls are made by stating the name of the procedure and a list of the objects whose values are to be passed to the procedure. The format is:

procedure_name (passing_parameter_list) ;

Since this is a statement, and not an expression, it must end with a semicolon and may not appear as an expression in another statement.

When the procedure call exists as a concurrent statement it is activated by changes in any signal associated with a parameter of mode **in** or **inout** in the passing parameter list. For example, consider the following statement:

find_min (array_1, out_1, old_1) ;

This statement "calls" the procedure which appeared in Illumination Box 11.4.5. Here is the first line in that procedure:

procedure find_min (**variable** a_ray : **in** int_array ;
signal min_val : **inout** INTEGER ;
signal old_min :**out** INTEGER) **is**

array_1 in the calling statement is a variable array (this is likely since the calling statement is a sequential statement and items appearing in the formal parameter list will be either signals, variables or constants). It is associated with a parameter *a_ray* of mode **in**. Therefore, the procedure statement activates when it is encountered in the sequential code.

Notice that this procedure could not be called from a concurrent statement area in the model because it contains a variable in the parameter list.

11.4.7 Passing Parameter List

- Specifies objects from, and to, which values pass
- Association by
  - Position
  - Name
- Values associated with in parameters are "protected from change"

11.4.7 Passing Parameter List

The parameter list in the procedure call statement is called the passing parameter list or formal parameter list. It specifies the objects from which values pass to the procedure. The two ways in which objects in this list can be associated with the objects in the procedure parameter list are the same as in function calls:

1. Positional Association

   In positional association the objects named in the procedure call expression are matched by position to the objects in the procedure parameter list. For example, consider the procedure whose first line is:

   ```
   procedure find_min (variable a_ray : in int_array ;
                       signal min_val : inout INTEGER;
                       signal old_min : out INTEGER) is
   ```

   and the procedure call

   ```
   find_min (array_1, out_1, old_1) ;
   ```

   The value in *array_1* would be associated with *a_ray*, the value *out_1* with *min_val* and the value *old_1* with *old_min* in the procedure, the association being done by position. *array_1* is the first object in the formal parameter list of the statement and *a_ray* is the first object in the parameter list of the procedure, and so on.

2. Named Association

   Named association matches the objects in the procedure call statement to the objects in the procedure parameter list by name. For example, a named call to the above procedure example would be:

   ```
   find_min (min_val => out_1, a_ray => array_1,
             old_min => old_1);
   ```

   In this example it is easy to see that *min_val* is associated with *out_1* and so on.

It is possible to mix positional association and named association in a procedure call statement. However, all positional associations must come first and once a named association is used all remaining associations must be made by name.

Note that it is never possible to have a greater number of objects in the formal parameter list than in the procedure parameter list.

11.5 Exercises

11.5.1 Write a model of a logical inverter using a procedure subprogram (NB. It is recognised that this would be an inefficient way to model an inverter. It is meant to be an exercise in the use of subprograms only).

11.5.2 Create a model of a one_bit_adder similar to that which appears in Section 9.9. In this model create a function or a procedure which will convert the Std_logic states to integer values. Use these integer values to create a sum. Then use the sum in a case statement or a selected signal assignment statement to determine the output signals.

11.5.3 Write subprogram models for the counter and comparator of the Scintillation Detector of Chapter 3

11.5.4 Write a description for a 4 bit comparator, starting with the one bit comparator of Exercise 6.10.6.

11.5.5 Starting with the one-bit-adder of 11.5.2 create a model for a 4-bit binary full adder.

11.5.6 Produce models to perform the following conversions:

- i) Binary to integer
- ii) Integer to binary
- iii) Std_logic_vector to integer
- iv) Integer to Std_logic_vector

Chapter 12

Types and Arithmetic Operations

Overview

This chapter is a "building" chapter. Depth is added to material presented in earlier chapters. Specifically, the concept of types and typing has been presented at an informal level in previous chapters. VHDL is a strongly typed language but the reader has been allowed to develop an intuitive feel for the use of types. In this chapter the use, and concept, of types is formalised. Because much of what can, and cannot, be done with arithmetic operations in VHDL is tied to types and typing arithmetic operations are also formally introduced in this chapter

In this chapter you will learn

- What is meant by types
- The purpose of types
- How types are used in modelling
- The effects of a strongly typed language

12.1 Typing

- type determines the values an object may assume and operations which may be performed upon it
- Aids in design verification
    - Data paths
    - Object values
- type Classifications
    - Scalar
    - Composite
    - File
    - Access
- Specific types declarations fall within these classifications

12.1 Typing

The type of an object determines two factors about the object:

1. The values an object may assume.
   As has been pointed out in previous chapters the declaration of a type identifies a value set associated with that type name. When an object is declared to be of a specific type it can only assume values of the value set of that type.
2. The operations which may be performed on the object.
   Operations such as logical operations and arithmetic operations are not automatically defined for given types. Except for certain predefined types which exists in the VHDL package STANDARD, how operators are applied to types must be declared by the creator, or user, of that type.

There are no specific types 'built-in' to VHDL. All object types must be declared. There are certain types which are said to be predefined because they appear in the VHDL package STANDARD. The contents of this package are specified by the IEEE and are automatically included in all VHDL models. These 'predefined' types were introduced in Section 5.3.3.

Typing aids in design verification by calling attention as early as possible to data paths which are not consistent and to object values which are not consistent.

If an object of a specific type is assigned to an object of another type an error is generated during compilation. This informs the modeller that an inconsistency exists in the model. Similarly, if a value which is not in the value set of the object's type is assigned to an object an error is generated. This indicates to the modeller that the intended use of the object has been violated.

Type declarations fall into specific classifications on the basis of the value set of the type:

1. Scalar
   Value sets composed into quantities arranged in the form of a scale.
2. Composite
   Groups of elements which contain scalar values.
3. File
   Value sets which are the contents of an external file.
4. Access
   Value sets which are pointers to objects of some other type.

12.1.1 Integer Types

- **"Integers" are the unbounded set of positive and negative whole numbers**
- **range guaranteed to include 32 bits**
- **Upper and lower range constraints must be integer numbers**
- **Declaration format**

  type type_name **is range** implementation_defined ;

- **Predefined integer type:**

  type INTEGER **is**
  range -2147483647 to 2147483647 ; --
  predefined numeric type

12.1.1 Integer Types

Scalar types can be formed from the unbound set of positive and negative whole numbers. These numbers are referred to as integer numbers. Note, however, that integer numbers should not be confused with the predefined INTEGER type. The type INTEGER declared in the VHDL package STANDARD is an integer type; however, other integer types can be declared. Integer types are scalar numbers where the upper and lower range constraints of type value set are integer numbers. The declaration format is:

type type_name **is range** implementation_defined ;

All type declarations must be made in the declarative areas of the model. If a type is used in a model or package the type declaration must precede the first use of that type. For example, the type Std_logic has been used in the **port** clause of the entity declaration to identify the type of the port. Since the type declaration for this type appears in the package Std_logic the **use** statement for Std_logic_logic must appear *before* the entity declaration. If it were to appear between the entity declaration and the architecture body the declaration for Std_logic would occur after the first use of the type.

The predefined INTEGER type may be declared as:

type INTEGER **is range** -2147483647 **to** 2147483647 ; -- predefined numeric type

It is obvious that there are lower and upper limits to this value set; therefore the INTEGER type is not an unbound set of positive and negative whole numbers. It is simply an integer type with the type name set to INTEGER. The range of INTEGER is implementation_dependent but it is guaranteed to include the above range (which is a range of 32 bits).

The type INTEGER is declared in the VHDL package STANDARD but since VHDL is implemented on many different types of hardware the range constraints for this type may vary from installation to installation but may not be smaller than that given above.

Note that it is permissible to create a type declaration in a model which has a type name of integer. For example:

type integer **is range** 0 **to** 100 ;

This would not be a good idea since the declaration of type integer will 'hide' the package STANDARD declaration in certain areas of the model but other areas may exist where the standard declaration is not hidden. This could produce confusing results whose source might be difficult to trace. As a general rule therefore, it is recommended that 'predefined' type names be not used in type declarations.

12.1.2 range

- **Identifies a subset of values**
- **May be used with type declarations or object declarations**
- **Format**

  range begin direction end

- **"Direction" may be:**

  Ascending - to
  Descending - downto

- **Examples**

type day **is range** 1 **to** 31 ;
type voltage **is range** 12 **downto** -12 ;
signal in_volts : voltage **range** 5 **downto** 0 ;

12.1.2 range

The **range** in a type declaration identifies the subset of values which identify the value set associated with that type. It may be used with type declarations involving numerical value sets such as integer and floating point types, or with object declarations where the value set for the object is a partial set of the value set of the type.

The format for **range** is:

range begining_value <direction> ending_value;

Two <directions> may be specified:

1. **to**

   to specifies an ascending range. Typically, the value to the left of **to** should be equal to or less than the value to the right of **to.** If the left-most value is made greater than the right-most a null value set will be created and the type rendered ineffective for identifying values.

2. **downto**

   downto specifies a descending range. Typically, the value to the left of **to** should be equal to or greater than the value to the right.

An example of a range which identifies an ascending integer value would be:

range 0 **to** 10 ;

and a descending floating point value set:

range 5.1 **downto** 1.1 ;

It is important to note that value on either side of the direction indicator cannot be mixed. An integer number cannot be used on one side and a floating point number on the other. The following example shows an invalid range structure:

range 0 **to** 5.1 ;

The use of **range** in an object declaration is to identify a partial set of values from within the original set of the type. The new range cannot be wholly, or partially, outside the range of the type. A valid example would be:

type whole_number **is range** 0 **to** 100 ;
signal example : whole_number **range** 0 **to** 10 ;

12.1.3 Floating Point Types

- **"Floating Points" are the unbounded set of positive and negative numbers which contain a decimal point**

- **range is restricted by limitation of host machine**

- **Upper and lower range constraints must contain a decimal point**

- **Declaration format**

  type type_name **is range** implementation_defined;

•**Predefined floating point type:**

type REAL **is range** -1.0E38 **to** 1.0E38 -- predefined numeric type

12.1.3 Floating Point Types

Floating point types can be formed from the unbound set of positive and negative numbers which contain a decimal point.

The type REAL declared in the VHDL package STANDARD is a floating point type. Unlike the predefined type INTEGER it is unlikely that type REAL will confuse the issue as to what is a floating point type and what is the type REAL. The format for declaring floating point types is similar to the format used when declaring integer types. The only difference is that the upper and lower range constraints of the type value set must be floating point numbers instead of integer numbers.

The format for floating point types is:

type type_name **is range** implementation_defined ;

All type declarations must be made in declarative areas of the model. If a type is used in a model or package the type declaration may precede the first use of that type.

The predefined type REAL is declared as:

type REAL **is range** -1.0E38 **to** 1.0E38; --predefined numeric type

The range of REAL is host-dependent but is guaranteed to include the above range. It is defined with an ascending range.

The following type declaration shows how a user-declared floating point type might be created:

type my_numbers **is range** 0.0 **to** 20.5 ;

Notice the use of the decimal point in the values. This identifies my_numbers as a floating point type. In this case the type has an ascending range. Descending ranges can be created using **downto**.

12.1.4 Enumeration Types

- **Lists of identifiers or character literals**
- **Identifiers follow standard naming rules**
- **Character literals are all of the upper and lower case alpha characters, numbers and special characters**
- **Declaration format**

  type type_name **is** (enumeration_ident_list) ;

- **Predefined enumeration types**

  type BIT **is** ('0', '1') ;
  type BOOLEAN **is** (FALSE, TRUE) ;
  type SEVERITY_LEVEL **is** (NOTE, WARNING,
  ERROR, FAILURE) ;
  type CHARACTER **is** ('a', 'b', 'c', . . .) ;

12.1.4 Enumeration Types

A third scalar type which can be identified is the enumeration type. This type is characterised by a value set consisting of identifiers or character literals.

1. Identifiers
   Identifiers are names. They follow the standard VHDL naming rules and may consist of any number of lower or upper case characters, numbers or underscores. Once a name is used as an identifier that name cannot be used as a signal, variable or constant name in an object declaration occurring at the same level as the type declaration. The name could appear as an object name if the object declaration occurred at a different level. However, when this happens the object name (or the identifier) is hidden by whichever of the declarations was made last.
2. Character literals
   Character literals are all upper and lower case alpha characters, numeric characters and most special characters.

   When character literals appear in an enumerated type value set they must appear within single quote marks.

   Although VHDL isgenerally case insensitive character literals are an exception to its case insensitive nature.

   Whenever a character literal is used in a model it must appear in single quotes and must have the same case as it had in the type declaration.

The type declaration format for enumeration types is:

type type_name **is** (enumeration_ident_list) ;

The numeration_ident_list must consist of identifiers, character literals or a mixture of identifiers and character literals.

Many of the predefined types in the VHDL STANDARD package are enumerated types. For example:

type BIT **is** ('0', '1') ;
type BOOLEAN **is** (FALSE, TRUE) ;
type SEVERITY_LEVEL **is** (NOTE, WARNING,
ERROR, FAILURE) ;

An interesting aspect of enumerated types is that they have a range of values. The left-most value in the set is the 'lowest' value, the right-most the 'highest' value. When an object is declared to be of an enumerated type it can assume a range within the range of the type. For example:

signal example : SEVERITY_LEVEL **range** NOTE **to** ERROR;

12.1.5 Physical Types

- **Describe objects in terms of a base unit, multiples of a base unit and a specified range**

- **Range restricted to integer values**

  range limits -2147483647 to 2147483647

- **Declaration format**

```
type type_name is range implementation_defined
    units
        fs ;               -- femtosecond
        ps = 1000 fs ;     -- picosecond
        ns = 1000 ps ;     -- nanosecond
        us = 1000 ns ;     -- microsecond
        ms = 1000 us ;     -- millisecond
        sec = 1000 ms ;    -- second
        min = 60 sec ;     -- minute
        hr = 60 min ;      -- hour
    end units ;
```

N.B. A separator is required between an abstract literal and the unit name

The use of this 'range' of values is that enumerated types can be used as the counter in a **for** loop. For example:

ex_loop : **for** i **in** NOTE **to** ERROR **loop** . . .

12.1.5 Physical Types

A fourth scalar type is the physical type. This type describes objects in terms of a base unit and a specified range. It also has the option of describing multiples of the base unit.

Values in the value set of a physical type must consist of a number and a unit.

The type declaration format for physical types is:

```
type type_name is range implementation_defined ;
   units
         base_unit ;
         [-- multiples of base unit ;]
   end units ;
```

Notice that the multiples of the base unit are optional.

The range_constraint in the format must be discrete units. Normally integer numbers are used for these constraints. For example:

```
type volume is range 0 to 100
   units
         cc ;
         dc = 10 cc ;
          L = 10 dc ;
   end units ;
```

The base unit is cc (cubic centimetres). A dc (decilitre) is the equivalent of 10 cc. A L (Litre) is the equivalent of 10 dc or 100 cc. All multiples of the base unit must be expressible in integer amounts of the base unit. When an object of type volume has a value assigned to it the value assigned must be expressed in terms of an amount <u>and</u> a unit. The 'value' 1.5 dc could be assigned to an object of type volume because it translates into 15 cc. On the other hand, 2.52 dc would be assigned as 25 cc. This is because all assignments to objects of physical types are in terms of a discrete number of base units. If the stated value does not translate into a discrete number of units it is rounded to the 'nearest' whole value.

N.B. a separator (space) is required between an abstract literal and its unit name.

12.1.6 Scalar Subtypes

- **Subsets of specified types**
- **Do not define new types**
- **range constraints must be within range of defining type**
- **Declaration format**

  subtype name **is** type_name **range** constraints ;

- **Predefined scalar (numeric) subtypes**

  subtype NATURAL **is** INTEGER **range**
  0 **to** 2147483647 ;

  subtype POSITIVE **is** INTEGER **range**
  1 **to** 2147483647 ;

12.1.6 Scalar Subtypes

Subtypes are subsets of specified types. They do not define new types. Their range constraints must be within the range of the defining type. Subtypes can be declared for all scalar types: numeric, enumeration and physical.

The general subtype declaration format is:

subtype subtype_name **is** type_name **range** range-constraint ;

Notice the use of the defining type name in the subtype declaration. The range which follows the type_name must be within the range of the declaring type. For example:

type int_number **is range** 0 **to** 100 ;
subtype sub_int **is** int_number **range** 10 **to** 50 ;

The range of the subtype is within the range of the type.

Subtypes allow the modeller to restrict the values an object may assume while allowing certain operations. For example:

type in_num **is range** 0 **to** 10 ;
type er_num **is range** 10 **to** 20 ;
signal a : in_num ;
signal b : er_num ;

Even though signals *a* and *b* are integer type numbers they cannot be added together. As will be seen later in this chapter, arithmetic operations are defined for objects of the same type. This is one of the aspects of a strongly typed language. However, the above example could be modified as follows:

type base_num **is range** 0 to 20 ;
subtype in_num **is** base_num **range** 0 **to** 10 ;
subtype er_num **is** base_num **range** 10 **to** 20 ;
signal a : in_num ;
signal b : er_num ;

Signals *a* and *b* can now be added together because they are subtypes of the same type. This ability provides the advantages of a strongly typed language without some of the disadvantages. Signals *a* and *b* are still checked for range violations in the desired value set but at the same time they can be used together in arithmetic operations.

There are two subtypes declared in the VHDL STANDARD package:

subtype NATURAL **is** INTEGER **range** 0 **to** 2147483647 ;
subtype POSITIVE **is** INTEGER **range** 1 **to** 2147483647 ;
-- predefined numeric subtypes.

12.1.7 array Types

- **Multiple values of the same type under a single identifier**
- **One or more dimensions**
- **Values referenced by indices**
- **Type of index must be integer or enumeration**
- **Declaration format**

  type array_type_name **is array**
  (positive **range** <>) **of** character ;

- **Predefined array types**

  type STRING **is array**
  (POSITIVE **range** <>) **of** CHARACTER ;

  type BIT_VECTOR **is array**
  (NATURAL **range** <>) **of** BIT ;

12.1.7 array Types

There are two kinds of array types: arrays and records. Array types identify a structure which can contain multiple values of the same type. The number of elements it contains, and the type of values which can be contained in these structures, is stated in the array declaration. Earlier chapters have introduced and used array objects and the reader may have developed an intuitive feel for what they are. The general type declaration format for arrays is:

type array_type_name **is array** (range_constraint) **of** <type>;

The range constraint which appears in this declaration can specify one or more dimensions. These values represent the indices used to reference the elements within an object of this type. These values must be discrete values. They can be either integer numbers, enumeration identifiers or enumeration character literals. For example, a type declaration involving an integer range would be:

type int_array **is array** (**range** 0 **to** 100) **of** INTEGER ;

This type declaration declares an array which consists of 101 elements that can hold values of type integer. The indices in objects of this type are the integer numbers 0 through 100.

A type declaration in which the range is defined in terms of enumerated identifiers would be:

type indices **is** (one, two, three, four, five, six) ;
type an_array **is array (range** one **to** five) **of** INTEGER ;

This type declaration declares an array type which consists of five elements which can hold values of type integer. To access an element in an object of this type the index would have to be given as one of the identifiers, viz: one, two, three, four or five.

The type of the element <type> is very important. It identifies the value set from which values can be chosen for storage in the elements of the array object. In many cases this type is a type declared within the model. If this is the case the element type must be declared before the array type is declared.

There are two array types declared in the package VHDL STANDARD:

type STRING **is array** (positive **range** <>**)** **of** CHARACTER;
type BIT_VECTOR **is array** (NATURAL **range** <>) **of** BIT;

12.2 array Range

- **Constrained or unconstrained**
- **Boundaries of constrained array are stated:**

```
type array_1 is array
        (Integer range -10 to 25) of BIT ;
type array_1_too is array (-10 to 25) of BIT ;
```

- **Boundaries of unconstrained array are left open**

```
type array_2 is array (INTEGER range <>) of BIT;
```

- **Boundaries can be enumerated types**

```
type pet is (dog, cat, bird, horse, kid) ;
type house_pet_array is array
            (pet range dog to bird) of INTEGER ;
type all_pets is array (pet range <>) of BIT ;
signal cage : all_pets(range cat to horse) ;
```

12.2 array Range

Array types can be classified into two categories based on the way in which the range of indices is expressed:

1. Constrained

   The indices boundaries of a constrained array are stated in terms of specific values. For example:

   type array_1 **is array** (-10 **to** 25) **of** BIT ;

   This declaration declares an array type which has thirty-six elements. These elements have index numbers -10 through 25 and can contain the values '0' or '1'. An object of this type can be declared as follows:

   variable example : array_1 ;

   This object has thirty-six elements which are addressed using the name example followed by a number -10 through 25 in parentheses. It is not necessary to assign a range constraint to an object declared to be of a constrained array type. However, the object can have a range constraint if this constraint is within the range of the constraint of the type.

2. Unconstrained

   The indices boundaries of an unconstrained array are stated in terms of a type but are left unqualified. For example:

   type array_2 **is array** (INTEGER **range** <>) **of** BIT ;

   This declaration declares an array type which has no limit on the number of elements. However, the values of the indices must be values which are in the set of values for the type INTEGER.

   Objects declared to be of an unconstrained array type must have a specific range stated when they are declared. For example:

   variable example : array_2 (**range** 0 **to** 25) ;

   This object has twenty-six elements which are addressed using the name example followed by a number 0 through 25 in parentheses. The range constraint is mandatory; it identifies the indices values and the number of elements in the array object.

12.2.1 Array Subtypes and Aliasing

- **Subsets of specified array types**
- **Do not define a new array type**
- **type that subtype is based upon must be an unconstrained array**
- **Declaration format**

```
subtype name is
        (array_name range range_constraint) ;
```

- **Example**

```
type data is array (NATURAL range <>) of BIT ;
subtype low_range is (data range 0 to 7) ;
subtype high_range is (data range 8 to 15) ;
```

12.2.1 array Subtypes and Aliasing

Just as there are subtypes associated with scalar types, there are subtypes associated with array subtypes. Array subtypes are subsets of specified array types. They have the same advantages as scalar subtypes in that they do not define new types.

The general subtype declaration format is:

subtype subtype_name **is**
(array_type **range** range_constraint) ;

Array subtypes can only be declared for unconstrained array types.

This is why the array_type must appear in the range constraint clause. For example:

type data **is array** (NATURAL **range <>) of** BIT ;
subtype low_range **is** (data **range** 0 **to** 7) ;
subtype high_range **is** (data **range** 8 **to** 15) ;

Object elements which are from objects of either low_range or high_range may be operated upon as if they were objects of the same type.

It is important to note that array subtypes can only be created for unconstrained arrays. To understand why this is so it must be realised that VHDL does not permit the creation of a subtype of a subtype. For example, the following is invalid:

subtype invalid **is** (low_range **range** 0 **to** 2); -- invalid

Invalid cannot be created as a subtype of low_range because low_range is already a subtype. The creation of subtypes of subtypes is not permitted in VHDL.

Subtypes of constrained arrays cannot be created because of the manner in which constrained arrays are created. Constrained arrays are created as subtypes. Therefore, since subtypes of subtypes cannot be created, subtypes of constrained arrays cannot be created.

An **alias** declaration declares an alternative name for an existing named entity. An object, a slice of an object, or an indexed part of an object can be given an alternative name by means of an alias declaration. This facility can become useful when subtypes or slices of arrays need to be named using discrete identifiers in preference to using indexed values of an array as, for example, in the partitioning of registers in microprocessors.

12.2.2 array Slices

- **Consecutive positions of one-dimensional arrays**

- **Created by reference**

- **Example**

```
process (a)
        type data is array (range 0 to 15) of BIT ;
        variable ray_a : data (0 to 15) ;
        variable ray_b : data ( 0 to 3) ;
begin
        ray_b := ray_a (4 to 7) ;
                    •
                    •
                    •
```

12.2.2 array Slices

Array slices are sections of array objects. They are created by reference and do not appear in any type of declaration. Consider the following code structure:

```
process (a)
    type data is array (range 0 to 15) of BIT ;
    variable ray_a : data (0 to 15) ;
    variable ray_b : data (0 to 3) ;
begin
    ray_b := ray_a (4 to 7) ;
```

The structure *ray_a (4 to 7)* in this example is an array slice. It represents only part of the elements in the array object *ray_a*.

Array slices are comprised of consecutive positions (elements) in a one dimensional array, viz:

1. Slices may only be taken from one dimensional array objects.
   Only one dimensional array types have been introduced and discussed in this text but the reader should be aware that multi-dimensional arrays can be created in VHDL. For example:
   ```
   type two_d_array is array (range 0 to 2, range 0 to 2) of BIT;
   ```
   declares a two dimensional array type which has nine elements.
2. Slices must consist of consecutive positions (elements) within the array.
   It is not possible to create a slice with non-consecutive elements. For example, the following is an invalid slice:
   ```
   ray_a (3, 5, 7, 9).
   ```
   However, it is possible to put the non-consecutive elements into an array using concatenation, viz:
   ```
   ray_b <= ray_a(3) & ray_a(5) & ray_a(7) & ray_a(9) ;
   ```

12.2.3 Predefined type Attributes

Most of the following predefined type attributes are valid with scalar types or subtypes. In some cases they are valid only with discrete types.

The definitions for the predefined type attributes listed below are given in Appendix A:
type'BASE, type'HIGH, type'LEFT, type'LEFTOF(x), type'LOW, type'POS(x), type'PRED(x), type'RIGHT, type'RIGHTOF(x), type'SUCC(x), type'VAL(x).

All of the above return the appropriate value appertaining to the value set of the type declaration to which they refer.

12.3 Strong Typing

- Objects may assume values only of their type
- Only operations defined for a type may be applied to objects of that type
- Strong typing captures intent and flags errors as early as possible

12.3 Strong Typing and Overloading

VHDL is a strongly typed language but, while strong typing captures the intent of the modeller and flags errors as early as possible in the modelling process, it does impose certain limitations, viz:

1. A value cannot be assigned to an object unless the value is part of the value set of the type of the object.
2. An object cannot be assigned to another object unless the objects are of the same type. Consider the following object declarations:

   signal a : Std_logic := 'X' ;
   signal b : BIT := '0' ;

   The following is an invalid assignment

   a <= b ; -- invalid assignment

   Even though a Std_logic signal has '0' as one of the bit values in its value set an attempt to assign a value from an object of one type to an object of another type results in an error. This is an aspect of strong typing which is designed to capture intent.
3. Only operations defined for a type may be applied to objects of that type.
   A type defines a set of values which an object may assume. It also defines a set of operations which can be performed on an object of that type.
4. Overloading.
   When an object is first declared there are no operations defined for that particular type. How operators are to be applied to objects of that type must be declared. This process is called "overloading the operator" for the particular type. In certain instances some operators are overloaded for all types. For example, relational operators are overloaded for objects of the same type and arithmetic operators are overloaded for any numeric object.

The remainder of this chapter introduces arithmetic operators and these are discussed in terms of how they function and are overloaded for numeric types.

12.4 Arithmetic Operations

- **Predefined binary operators**

Symbol	Function
**	exponentiation
*	multiplication
/	division
mod	modulus
rem	remainder
+	addition
-	subtraction

- **Predefined unary operators**

Symbol	Function
+	plus
-	minus
abs	absolute value

12.4 Arithmetic Operations

There are three classes of operators in VHDL; logical operators, relational operators and arithmetic operators. Logical and relational operators have already been introduced and discussed (Sections 5.4 and 8.3). The following discourse is confined to the introduction and discussion of arithmetic operators.

Arithmetic operators in VHDL are classified by the number of operands in the operation; these impose certain limitations:

1. Binary operators

   Binary operators require two operands in the following format:

   left_operand operator right_operand

   There are seven predefined binary operators:

Symbol	Function
**	exponentiation
*	multiplication
/	division
mod	modulus
rem	remainder
+	addition
-	subtraction

2. Unary operators

   Unary operators require one operand in the following format:

   operator right_operand

   There are three predefined unary operators:

Symbol	Function
+	plus
-	minus
abs	absolute value

12.4.1 ** (exponentiation operation)

- **Raises objects to powers (exponentiation)**
- **Format**

  left_operand ** right_operand

- **Left_operand**

  integer or floating point object/number

- **Right_operand**

  integer object/number

- **Result**

  same as left_operand

- **Negative right_operands used only with floating point left_operands**

12.4.1 ** (exponentiation operation)

The double asterisk (**) is used as an exponentiation operation to raise numeric values to powers. It is a binary operator and requires two operands in the following format:

left_operand ** right_operand

Several factors must be considered when choosing either numbers or objects for use with the power operator:

1. The left operand may be an object or number which is of integer or floating point type. So long as the type of the object is either integer or floating point based the object name may be used as the left operand. This includes both positive and negative values.
2. The right operand must be an integer number or an object of type integer. Floating point values of objects or floating point type may not be used as the right operand. This may produce an interesting quandary for modellers accustomed to determining the roots of numbers by raising them to fractional powers. For example, the square root of 4 could be found by raising it to the power of 0.5, i.e. 2 = 4**0.5. This is <u>not</u> valid in VHDL (which is in common with a number of software languages such as ADA). Modellers in VHDL must create routines to find the roots of numbers, or use vendor supplied mathematics packages.
3. The result of an exponentiation operation has the same type as the left operand. If the left operand is an integer type or value the result is an integer value. If it is a floating point type or value the result is a floating point value. This is important when considering the type of object to assign the result to, or when the result is used in addition arithmetic computations.
4. Negative right operands can be used only with floating point left operands. Raising a value to a negative power is the same as raising the value to a positive power and calculating its reciprocal i.e. dividing 1 by the result e.g., 2**-2 = 1/(2**2). The result will always be a decimal fraction. Since Rule 3 demands that the left operand and the result be of the same type only floating point left operands may be use with negative powers. Note that the form 2**-2 is an illegal form of the expression in VHDL; it must be written as 2**(-2) (See Arithmetic precedence, Section 12.6). However, 1/2**2 would be legal and have the meaning intended above.
5. Raising any value to the power zero results in 1.
6. Raising a floating point value to a power produces an approximate result.

12.4.2 * and / Operators

- **Perform multiplication (*) and division (/)**
- **Format**

  left_operand * right_operand
  left_operand / right_operand

- **Result type**

left operand	right operand	result
integer	integer	integer
	physical	physical
floating pt	floating pt	floating pt
	physical	physical
physical	integer	physical
	floating pt	physical
(division only)		
physical	physical	integer

12.4.2 * and / Operators

The single asterisk(*) operator is used to perform multiplication and the slash (/) operator is used to perform division. Both of these operators are binary operators and require two operands in the following format:

left_operand * right_operand
left_operand / right_operand

Floating point, integer and physical values and types may be used as either the left or right operand, under certain conditions. The following table shows where and when these values can be used.

Multiplication and Division

left operand	right operand	result
integer	integer	integer
	physical	physical
floating point	floating point	floating point
	physical	physical
physical	integer	physical
	floating point	physical
(division only)		
physical	physical	integer

Notice that the type of the left operand determines the type that the right operand can assume, and the type of the result. In all cases the right operand must be the same type as the left operand, or a physical type. The result type is the same as the left operand type except when a physical type is divided by a physical type. In this case an integer results because the units cancel, e.g. cc/cc = 1, dc/cc = 10 (cf.. Section 12.1.5). The result, although dimensionless, nevertheless has a relationship with volume.

Note that a physical type cannot be multiplied by a physical type.

12.4.3 mod and rem

- **Remainder operators defined only for integer types**

- **Format**
  left_operand mod right_operand
  left_operand rem right_operand

- **rem results**
  left_operand positive:
  positive remainder of
  abs(left_operand) / abs(right_operand)
  left_operand negative:
  negative remainder of
  abs(left_operand) / abs(right_operand)

- **mod results**
  left and right_operand have the same sign:
  remainder of abs(left_operand) divided by abs(right_operand) with sign of right_operand

  left and right_operand have different signs:
  a value less than abs(right_operand) with the sign of right operand such that
  left_operand = right_operand*integer + value

12.4.3 mod and rem

mod and **rem** are remainder operators. The result produced by these operators is the remainder of a "division-like" operation within modulo n arithmetic, where n is the divisor. **mod** and **rem** are defined only for integer type objects and integer values. Both of these operators are binary operators and require two operands in the following format:

left_operand **mod** right_operand
left_operand **rem** right_operand

1. **rem**

   This operator produces a value which is the remainder of the left operand divided by the right operand and which has the sign of the left operand. For example, the result of:

   5 **rem** 2 is the remainder of 5/2 with the sign of 5, i.e. 1
   8 **rem** -3 is the remainder of 8/3 with the sign of 8, i.e. 2
   -9 **rem** 2 is the remainder of 9/2 with the sign of -9, i.e. -1
   -11 **rem** -4 is the remainder of 11/4 with the sign of -11, i.e. -3

   Note that the division is always performed on the absolute values of the two operands and that the sign of the result is always the sign of the left operand.

2. **mod**

   This operator produces a value which depends upon the sign of the left operand.

   If the left and right operands have the same sign the value produced is the remainder of the left operand divided by the right operand taking the sign of the right operand.

   If the operands have different signs the value produced is such that

   left_operand = right_operand*integer + value
   or A = B*N + (A **mod** B)

   the value being less than **abs**(right_operand) and taking the sign of the right operand. For example, the result (value) of:

   10 **mod** 3 is the remainder of 10/3 with the sign of 3, i.e. 1
   10 **mod** -3 is the remainder of 10 - (3*4) with the sign of -3, i.e.-2
   -10 **mod** -3 is the remainder of 10/3 with the sign of -3, i.e. -1
   -10 **mod** 3 is the remainder of 10 - (3*4) with the sign of 3, i.e. 2

   Note that the subtraction is always performed on the absolute values of the two operands and that the sign of the result is always the sign of the right operand. Note also that the integer in the expression is that multiplier which will cause the product of the right operand and the integer to exceed the absolute value of the left operand by less than the absolute value of the right operand.

12.4.4 Binary and Unary + and -; abs

- **Binary + and -**
  - Perform addition (+) and subtraction (-)
  - Format
    left_operand + right_operand
    left_operand - right_operand
  - Left_operand
    integer, floating point or physical object/value
  - Right_operand
    same type as left_operand
  - Result:
    same type as left_operand
- **Unary + and -**
  - Sign operators
    + (identity function)
    - (negation function)
  - Format
    + right_operand
    - right_operand
  - Right_operand
    integer, floating point or physical object/value
  - Result
    same type as right operand
    N.B. Cannot appear immediately before or after binary

  arithmetic operator
- **abs**
  - Returns positive value of operand
  - Format
    abs(right_operand)
  - Right_operand
    integer, floating point or physical object/value
  - Result
    same type as right operand

12.4.4 Binary and Unary + and - ; abs

1. Binary + and -

   The plus (+) operator is used to perform addition and the minus (-) operator to perform subtraction. These binary operators require two operands in the following format:

   left_operand + right_operand
   left_operand - right_operand

   Floating point, integer and physical values and objects may be used as the left operand. Once a specific object or value is chosen for the left operand, the right operand must be of the same type.

   The result of either an addition or subtraction is always the same type as the left operand.

2. Unary + and -

   The unary operators + and - are defined as 'sign' operators and must not be confused with the addition and subtraction operators. These operators represent the identity function (+) and the negation function(-). As unary operators they require only one operand viz:

   \+ right_operand
   \- right_operand

   The negation function is most often used to change the sign of a value, but it is important to recognise that neither the negation function nor the identity function can appear immediately before, or after, a binary arithmetic operator. For example:

   a*-b is illegal since the negation of b occurs immediately after the multiplication operator.

   a*(-b) is legal since the negation of b occurs as a separate process

   -a*b is legal since the negation of a does not occur immediately before the multiplication operator.

   a-*b is illegal since the presumed intention to negate b occurs immediately before the multiplication operator

3. **abs**

   abs is a unary operator which returns the absolute value of the operand that is to its right. When it is used the operand must appear in parentheses:

   abs (right_operand)

   The operand may be any integer, floating point or physical object or value. The result will be the same type as the operand.

12.5 subtype Calculations

- **Different subtypes of same type do not violate strong typing conventions**
- **Example**

```
power : process (clk)
        type num is range  1 to 20 ;
        subtype  a is num range 1 to 10 ;
        subtype b is num range 10 to 20 ;
        variable x : a ;
        variable y, z : b ;
    begin
        z := x + y ;
    end process power ;
```

12.5 subtype Calculations

One aspect of binary arithmetic operators is that they are overloaded for all numeric types if the same type is used for both operands or a value of that type is used. This does no violate the purpose of a strongly typed language such as VHDL. Two objects of different numeric types still cannot be added together (nor any other arithmetic operation). This is true even if the types are both integer or both floating point.

Consider the following example, which is of an illegal addition:

```
type_example :  process (clk)
                    type a is range 1 to 10 ;
                    type b is range 10 to 20 ;
                    variable x : a ;
                    variable y, z : b ;
                begin
                    z := x + y ; -- illegal
                end process type_example ;
```

The addition of x and y is illegal because the right and left operands are of different integer types. If the left operand is an object of a specific integer type, the right operand must be of exactly the same type. Both objects in the above example contain integer values but the fact that they are of different types prevents their addition. Similar problems will occur with other arithmetic operators.

The above problem may be solved in one of two ways:

1. Both objects could be declared to be the same type.

```
type_example : process (clk)
                    type  a is range 1 to 20 ;
                    variable x, y, z : a ; -- legal
                begin
                    z := x + y ;  -- legal
                end process type_example ;
```

2. The objects could be declared to be subtypes of the same type.

```
type_example : process (clk)
                    type num is range 1 to 20 ;
                    subtype a is num range 1 to 10 ;
                    subtype b is num range 10 to 20 ;
                    variable x : a ;
                    variable y, z : b ;
                begin
                    z := x + y ; -- legal
                end process type_example ;
```

Using different subtypes of the same type does not violate strong typing.

12.6 Arithmetic Precedence

- **Order in which calculations are performed**
- **Expressions are evaluated left to right in the following order**

Symbol	Evaluated
**, abs	first pass
*, /, mod, rem	second pass
sign (unary +, -)	third pass
+, -	fourth pass

12.6 Arithmetic Precedence

The three classes of operator have now been introduced; arithmetic, relational and logical. These operators are used in expressions. Many times only one class of operator is used in an expression. Consider the following **if** statement:

if a < b **and** c > d **then . . .**

The expression *a* < *b* ***and*** *c* > *d* mixes relational operators and logical operators. How will such an expression be evaluated? Is the result of *a* < *b* **and**ed with *c* or is *b* **and**ed with *c* and the result checked to see if it is greater than *a* and *d*?

To prevent chaos in such situations operators are evaluated in a specific order called the precedence of evaluation. When expressions are written which contain more than one class of operator the order of precedence is as follows:

All arithmetic operators are evaluated first;
secondly, all relational operators;
finally, all logical operators.

In the above example, therefore, *a* < *b* and *c* > *d* are evaluated first. The results of these two relational operations are then **and**ed together to achieve the final result.

Just as there must also be some order of precedence for the evaluation of classes of operators there must be a precedence for the evaluation of the arithmetic operators in arithmetic expressions. This order of precedence is shown in Illumination Box 12.6

When an arithmetic expression is evaluated the expression is examined for operators in each of the following classes:

1. ** and **abs**
2. *. /, **mod** and **rem**
3. sign (unary + and -)
4. + and -

Class 1 has the highest precedence. if an operator in this class is encountered in the expression it is performed before an operator in any other class. If no class 1 operators are encountered class 2 operators are evaluated in the order in which they appear in the expression. Then class 3 operators are evaluated and so on. The results of each evaluation are used in the next evaluation. Hence:

10*2 + 8/2 -1

evaluates to 23. If the result required is 49, 59, 13 or 28, which are the other possibilities using different precedents, then the appropriate parentheses must be inserted e.g.

10*(2 + 8/2) -1 = 59
10*(2 + 8)/2 -1 = 49

parentheses taking precedence over all other operators.

12.7 Exercises

12.7.1 Create a model of a device that 'smooths' certain waveforms on a Std_logic line. This device 'watches' the input line for 100 ns and then outputs on the output line the waveform which occurred the most number of times during the 100 ns. The model should record the number of 'X's, '0's, '1's and 'Z's which occur in a 100 ns period. At the end of the sampling period it must output, on another line, the state which occurred the greatest number of times.

12.7.2 For the IEEE supplied package Std_logic_1164 produce the type declarations and function declarations necessary to declare the overloaded NAND operator of Exercise 5.6.9.

12.7.3 Write a procedure which converts an integer value into its equivalent Std_logic_vector value.

12.7.4 Write a procedure which converts a Std_logic_vector value into its equivalent integer value.

12.7.5 Define the physical type resistance over the range milli-ohm to giga-ohm and capacitance over the range femtofarad to farad. Assume the host machine has a 32 bit limitation and that the parameters can take only positive values.

12.7.6 Using the types defined in Exercise 12.7.5 to define the rise and fall time constants for the R-C networks listed below.

t_rise : R = 1.0 Megohm C = 10.0 femtofarad
t_fall : R = 0.5 Megohm C = 10.0 femtofarad

Chapter 13

Type Conversions, Overloading and Resolution Functions

Overview

This chapter discusses methods for resolving problems associated with modelling in VHDL. A strongly typed language offers the user certain advantages and disadvantages. It captures intent and flags errors as early as possible in the design process. But it does mean, for example, that integer and floating point numbers cannot be added together, logical and relational operators cannot be applied directly to operands of different types and logical operators cannot be applied to any user declared type. This chapter discusses how some of these disadvantages can be overcome.

In this chapter you will learn

- How to overcome the disadvantages of strong typing
- The processes associated with overloading
- The concept of resolution functions

13.1 Converting Types

- **Changing values from one type to another**

- **Strong typing does not allow implicit value conversions**

- **Example of illegal usage**

```
process (clk)
        variable int: INTEGER range 0 to 10 ;
        variable flt : REAL range 0.0 to 10.0 ;
begin
        flt := int ; --illegal
end process ;
```

- **Explicit conversion is necessary**

    - **typemark conversion**
    - **user created conversion**

13.1 Converting Types

A strongly typed language offers the user certain advantages at the expense of some disadvantages. A major disadvantage is that types cannot be mixed in arithmetic operations. For example, two integer values cannot be added together if they are stored in objects of different type Similarly, integer and real numbers cannot be mixed in arithmetic operations.

Study the following two processes:

```
illegal_1 : process (clk)
        type type_1 is INTEGER range 0 to 10 ;
        type type_2 is INTEGER range 0 to 20 ;
        variable int_1 : type_1;
        variable int_2, int_3 : type_2
    begin
        int_3 := int_1 + int_2 -- this is illegal
    end process illegal_1 ;

illegal_2 : process (clk)
        variable int : INTEGER range 0 to 10 ;
        variable flt : REAL range 0.0 to 10.0 ;
    begin
        flt := int ; -- this is illegal
    end process illegal_2 ;
```

Even though *int_1* and *int_2* in illegal_1 are objects containing integer values they cannot be added together. Strong typing prevents arithmetic operations on objects of different types (unless the operator has been user overloaded for the particular types). In example illegal_2 an attempt is made implicitly to convert the integer value in int to a floating point number by assigning its value to a floating point object. This is also illegal.

These problems could be overcome if the values in an object of one type could be changed to another type. This would involve an explicit conversion of value type. In VHDL such conversions can be done in two ways:

1. Typemark conversion
   Typemark conversion allows the conversion of one closely related type into another. In example illegal_1 the types are both integer. It would seem reasonable that a simple conversion would allow the objects to be added together.
2. User created functions
   It is always possible to create a function to return any type value based on the passing parameter. Therefore, functions could be used as conversion mechanisms.

13.1.1 Typemark Conversion

- **Converts from one closely related type to another**

- **Format**

  typemark (object)

- **Typemark**

  name of type to be converted to

- **Example**

```
process (clk)
    variable int_er : INTEGER range 0 to 10 ;
    variable flt_er : REAL range 0.0 to 10.0 ;
begin
    tot := REAL (int_er) + flt_er ;
end process ;
```

13.1.1 Typemark Conversion

Typemark conversion allows the values of closely related types to be changed from one type to another.

Conversions of this type are explicit conversions and are allowed for scalar and array types. A typemark is simply the name of the type to which the value of the object is to be converted. Typemark conversions are made using the typemark and the name of the object:

typemark (object_name)

The example, illegal_1, which appeared in Illumination Box 13.1 could be made functional by typemark conversion:

```
legal_1 : process (clk)
        type type_1 is INTEGER range 0 to 10 ;
        type type_2 is INTEGER range 0 to 20 ;
        variable int_1 : type_1 ;
        variable int_2, int_3 : type_2 ;
    begin
        int_3 := type_2(int_1) + int_2 ; -- this is legal
    end process legal_1 ;
```

The problem of floating point values and integer values can be handled in the same way:

```
legal_2 : process (clk)
        variable int_er : INTEGER range 0 to 10 ;
        variable flt_er : REAL range 0.0 to 10.0 ;
    begin
        tot := REAL(int_er) + flt_er ; -- this is legal
    end process legal_2 ;
```

13.1.2 Valid Typemark Conversions

- **Closely related types**

From	To
Integer	Integer
Floating point	Floating point
Integer	Floating point
Floating point	Integer

- **Example**

```
process (clk)
        type a is REAL range 10.0 to 12.0 ;
        type b is REAL range 10.0 to 30.0 ;
        variable hat : a ;
                pat : b range 10.0 to 15.0 ;
                tot : b ;
begin
        tot := b(hat) + pat ;
end process ;
```

13.1.2 Valid Typemark Conversions

Typemark conversions involve the use of the name of the type and a parenthetical operand.

type_name (operand)

The operand can be any valid expression. It is usually the name of the object containing the value that is to be converted. Valid typemark conversions include:

1. Conversions between scalar numeric types.

   As indicated in the table in Illumination Box 13.1.2, the operand can be any integer or floating point type so long as the value converted to is also an integer or floating point type. Integer values may be converted to floating point types or to other integer types. Floating point types may be converted to integer types or to other floating point types.

   Typemark conversion does not negate constraints imposed by the type. For example, objects often have range constraints imposed upon them. Typemark conversion cannot overcome these constraints:

```
illegal : process (clk)
              type type_1 is INTEGER range 0 to 10 ;
              type type_2 is INTEGER range 11 to 20 ;
              variable int_1 : type_1 ;
              variable int_2, int_3 : type_2 ;
          begin
              int_3 := type_2(int_1) + int_2 ; -- error
          end process illegal ;
```

   This code segment causes an error either during compile or during execution. The value in the object *int_1* cannot be converted to type_2 because it does not fall within the range of type_2 values.

   Conversion of a floating point value to an integer value rounds the value to the nearest integer. Values that are exactly halfway between two integers are rounded up to the next integer. This means that 3.5 would be rounded to 4 and so on.

2. Conversion between array types.

   Array types can be used as typemarks provided the number of elements in each dimension of the arrays is the same, the array indices are of the same type or can be converted to the same type and the elements contain values of the same type.

13.1.3 User-created Conversions

- **Functions created by user**
- **Offer wider range of conversion possibilities**
- **Example**

```
function int_to_flt (in_val : range 1 to 4)
                                    return REAL is
begin
        case in_val is
                when 1=> return 1.0 ;
                when 2=> return 2.0 ;
                when 3=> return 3.0 ;
                when 4=> return 4.0 ;
        end case ;
end int_to_flt ;
```

- **Calling**

```
float_sum <= int_to_flt (int_val) + float_val ;
```

13.1.3 User-created Conversions

Functions have the ability to return values of any type. When the function is declared the type of the value returned is also declared. This feature makes it possible to create user defined conversion functions. For example:

```
function int_to_flt (in_val : range 1 to 4) return REAL is
  begin
        case in_val is
              when 1 => return 1.0 ;
              when 2 => return 2.0 ;
              when 3 => return 3.0 ;
              when 4 => return 4.0 ;
        end case ;
  end int_to_flt ;
```

This function could be used to convert the integer values 1, 2, 3 and 4 to floating point values. It would be called in the same manner as any other function:

```
float_sum <= int_to_flt (int_val) + float_val ;
```

The advantage of user created conversion functions is that they offer a wider range of conversion possibilities. Any type may be converted to another type. In addition, because functions only return values there is no range checking.

13.2 Overloading

- Giving more than one meaning to an item
- Needed when
  - same enumeration identifier appears in two or more enumeration type declarations
  - a single subprogram name addresses multiple subprograms
  - standard operators are not defined for user-defined types
- Overloading may be applied to
  - enumeration identifiers
  - names of functions and procedures
  - operator symbols
- Object names may be hidden but not overloaded

13.2 Overloading

There is a third way in which arithmetic operators can operate on objects of different types. The functionality of the operator itself can be changed.

Overloading is the process of giving more than one meaning to an item. The name John Smith is "overloaded" in the sense that it applies to many individuals. Such overloading is needed in VHDL when:

1. The same enumeration identifier appears in two or more enumeration type declarations.
2. A single subprogram name is used to identify several different subprograms.
3. Standard operators are not defined for user defined types, or the user wishes to change how an operator is applied to a particular type.

For these reasons overloading in VHDL may be applied to:

1. Enumeration identifiers.
2. Names of functions and procedures.
3. Operator symbols.

13.2.1 Enumeration Overloading

- **use of the same identifier in identifier list of two or more enumeration type declarations**
- **Allows conventional values to be used in non-ambiguous fashion**
- **Example**

type motorcycle **is** (triumph, bsa, ajs, vincent) ;
type car **is** (triumph, mg, austin, rover) ;

type count_cnt **is** (load, clear, accumulate) ;
type reg_cnt **is** (hold, clear, load) ;

13.2.1 Enumeration Overloading

Enumeration overloading usually refers to the use of the same identifier name or character literal in two or more enumeration type declarations. For example:

type motorcycle **is** (triumph, bsa, ajs, vincent) ;
type car **is** (triumph, mg, austin, rover) ;

or

type count_control **is** (load, clear, accumulate) ;
type register_control **is** (hold, clear, load) ;

In the first example, triumph appears in both identifier lists. In the second, clear and load appears in both lists. Overloading allows these names to be used in unambiguous fashions.

Consider the following assignment statement:

a_signal = '1' **when** in_put = 'X' **else**
'0' ;

Does the value '1' represent a value of type BIT or a value of type Std_logic? '1' is a value in both of these types and could, conceivably, cause some confusion for VHDL. However, because enumeration identifier lists are overloaded there is no problem identifying which list this '1' is from.

The user does not have to take any action with regard to the overloading of values appearing in enumeration identifier lists. VHDL automatically applies overloading principles to values appearing in these lists. There are, however, certain limitations.

Object names cannot be overloaded with enumeration identifiers. If an object declaration name appears at the same level as an enumerated type declaration that uses the object name as an identifier, an error results. If the object declaration and enumerated type declaration appear at different levels, one hides the other.

Enumeration identifiers can be overloaded for other enumeration identifiers, or for subprogram names.

13.2.2 Subprogram Overloading

- **Use of same name for two or more subprograms**

- **Allows application of subprogram functionality to parameter lists of differing type**

- **Example**

```
function min (a, b : INTEGER)
                    return INTEGER is
begin
    if a< b then return a ;
        else return b ;
    end if ;
end min ;

function min (fa,fb : REAL)
                    return REAL is
begin
    if fa < fb then return fa ;
        else return fb ;
    end if ;
end min ;
```

13.2.2 Subprogram Overloading

Subprograms may also be overloaded. Overloading in this case refers to the use of the same name for two or more subprograms. Because subprogram names may be overloaded a single name may be used to select the functionality of several different subprograms.

Consider the two functions in the example given in Illumination Box 13.2.2. Both functions have the same name, *min*. Both perform the same functionality. One accepts two integer values and returns the smaller of the two and the other accepts two real values and returns the smaller of the two.

Because of the restrictions of a strongly typed language and because functions can return a value of only one type, one function could not be written to return the minimum of both integer and floating point numbers. On the other hand, the user does not want to have several very similar yet slightly different names for different functions with the same basic functionality. Overloading allows the same name to be used in an unambiguous fashion. In this case the name min can be used to pick the smaller of two values whether those values are integers or floating point numbers.

Obviously when one name appears multiple times there must be some way for the system to differentiate among the various functions. In this discussion it has been implied that the parameter lists have something to do with this differentiation. This is part of the differentiation process but it is not the only factor involved in the selection of the proper subprogram.

13.2.3 Overloaded Subprogram Calls

- **Subprogram selection**
  - **number of parameters appearing in call**
  - **types of parameters appearing in call**
  - **names of parameters (if call uses named parameter associations)**
  - **in functions, the type of return value**

- **Examples**

```
(1)  function example (ix, iy : INTEGER)
                    return INTEGER is . .
(2)  function example (fx, fy : REAL)
                    return INTEGER is . .

(3)  function example (fx2, fy2 : REAL)
                    return REAL is . .

(4)  function example (a, b, c : INTEGER)
                    return INTEGER is . .

(5)  function example (aix : INTEGER ; biy : REAL)
                    return INTEGER is . .

     pdq := example (first_in, second_in) ;
```

13.2.3 Overloaded Subprogram Calls

In situations where there are several subprograms with the same name there must be sòme criteria for selecting the appropriate subprogram for a given calling situation. These selection criteria are listed in hierarchical order below:

1. Number of parameters appearing in call
   If two or more subprograms have the same name the system begins its selection process by eliminating all subprograms which do not have the same number of parameters in the calling list and the receiving list.
2. Types of parameters appearing in call
   If all but one of the subprograms cannot be eliminated by the number of parameters in the parameter lists, the types of the parameters in the two lists are compared. Because objects of like type must appear in both lists for values to be passed (strong typing) the proper subprogram can often be picked by type comparison.
3. Names of parameters if call uses named parameter associations.
   In those cases where subprogram calls are made by named association the proper subprogram can sometimes be picked by use of the parameter name in the calling statement.
4. In functions, the type of the return value.
   Because functions have a return value of a specific type this value can sometimes be picked by use of the parameter name in the calling statement.

Consider the example in Illumination Box 13.2.3. There are five functions named *example* declared in the model. A call to *example* is made. Which function is executed?

Since the call has only two parameters in its list, function *number 4* can be eliminated because the passing parameter list in this function requires three parameters. This still leaves four possibilities, so the second selection criterion must be applied.

Suppose that both of the objects in the calling list are objects of type real. This would eliminate functions from *numbers 1* and *5* from consideration. The objects in their lists are either integer or integer and real mixed.

Only functions *2* and *3* remain. Since the call is not made by named association, selection criteria 3 is of no use.

Criterion 4 looks at the return value. Suppose that the variable *pdq* is type integer. For the correct assignment to be made the return value must be type integer. This eliminates function *number 3* and so function *number 2* is the proper selection.

13.2.4 Overloading Operators

- **Existing operator symbols may be overloaded for use with user-defined types**
- **Logical, relational and arithmetic operators are really functions**
- **Follow same rules as overloading functions**
- **Function name is operator symbol in double quotes**

  function "operator_symbol" (parameter_list) . . .
- **Example**

```
function "+" (in_1 : new_state : in_2 : pe_op)
                                    return new_state is
begin
      if in_1 = X then
            if in_2 = '1' then return '1' ;
            elsif in_2 = '2' then return 'U' ;
                  •
                  •
                  •
end "+" ;
```

13.2.4 Overloading Operators

In addition to being able to overload enumeration identifiers and subprograms operator symbols may be overloaded. In essence, logical, relational and arithmetic are functions. The operator symbol is the name of the function and the rules for overloading the operator are the same as the rules for overloading functions.

All of the following predefined operators may be overloaded:

1. Arithmetic operators
   operator symbols: ****, abs, not, *, /, mod, rem, +, -, &**
2. Relational operators
   operator symbols: =, /=, <, <=, >, >=
3. Logical operators
   operator symbols: **and, or, nand, nor, xor, xnor**

The symbols "+" and "-" can be overloaded as both a binary operator and as a unary operator.

The actual overloading of the symbol is done by creating a function body (and declaration if necessary) in which the name of the function is the operator symbol in double quotes. The format would be:

```
function "operator_symbol" (passing_parameter_list) return . . .
```

The operator symbol must appear in double quotes and the only symbols which can be overloaded in this manner are those given above.

An example of a symbol overload function would be:

```
function "OR" (a: my_state; b: you_state) return our_state is
begin
if a = 'CA' then return 'CA' ;
elsif b = 'CA' then return 'CA' ;
else return 'XX' ;
end if ;
end "OR" ;
```

The example in Illumination Box 13.2.4 shows a similar overloading of the (+) operator to add objects of type *new_state* and *pe_op*.

13.2.5 Overloaded Operator Calls

- **Styles**
  - **function notation**
  - **operator notation**
- **Example of function notation**

```
x := "+" (y, z ) ;
```

- **Example of operator notation**

```
x := y + z ;
```

- **Determination of which overload is used for an operator follows overloaded subprogram rules**

13.2.5 Overloaded Operator Calls

The use of overloaded operator symbols offers two calling possibilities:

1. Calling by function notation

   Functions are called by using the name of the function in front of a parenthetical list of the object names whose values are to be passed to the function. Overloaded operator symbols can be called in the same fashion:

   ```
   x := "+" (y, z) ;
   ```

   This expression shows how the overloaded plus operator could be called. In this particular case the values of *y* and *x* are passed to the function named "+".

   The plus operator is a binary operator. Hence the object *y* corresponds to the left operand and the object *z* corresponds to the right operand.

2. Calling by operator notation

   Overloaded operator symbols can also be called by operator notation. In this form the operator is used in its standard fashion. For example:

   ```
   x := y + z ;
   ```

   produces the same effect as calling the overloaded plus operator in function notation.

   The left operand is passed to the first parameter in the function parameter list and the right operand is passed to the second parameter in the list.

When an operator symbol is overloaded several times the determination of which function to use in a particular case follows the same rules as those discussed for overloaded subprograms (see Illumination Box 13.2.3).

13.2.6 Operator Overloading Rules

- Only existing operator symbols may appear as function names
- Unary operators must have only one parameter
- Binary operators must have two parameters
- First list parameter must correspond to left operand; second operator to right operand
- Overloading does not affect level of hierarchy

13.2.6 Operator Overloading Rules

Certain rules must be followed when overloading operators. These rules have already been alluded to during the discussion of operator overloading. The following is a formal presentation of the rules:

1. Only existing operator symbols may appear as function names.
   Functions can be named anything the user wishes as long as the names follow the standard VHDL naming conventions. A majority of the operator symbols do not follow the standard naming conventions. Therefore the use of special symbols as function names is limited to the existing (predefined) operator symbols.
2. Unary operators must have only one parameter in the overload function.
   This is part of a more general rule. The number of parameters for an overloaded function operator must match the number of operands used by the operator.
   This rule is important when it is remembered that overloaded operators can be called using operator notation. This notation restricts the number of parameters to the number of operands. There can be only one or two operands; therefore there can be only one or two parameters.
3. Binary operators must have two parameters in the overload function.
   This is also part of the general rule discussed in 2 above. Binary operators must have two operands; therefore the function must have exactly two parameters.
4. The first list parameter must correspond with the left operand and the second parameter with the right operand.
   This is also part of the rule discussed in 2 above. It is needed in order to avoid confusion when the function is called in operator notation.
5. The "=" cannot be overloaded for **case** statements or selected signal assignment statements.
   This refers to the implied equality operation which is performed in the **case** statement. Each **when** clause in the **case** statement implies an equal comparison between the object stated after the **case** and the value stated after **when**.
6. Overloading does not affect the level of hierarchy for operators.
   When an expression is evaluated all arithmetic operators are evaluated first, then all relational operators and finally all logical operators. The overloading of operators does not affect this order of evaluation.

13.3 Multiple Signal Drivers

- Signal drivers are created when signal assignment statements are executed
- Drivers contain projected output waveforms
- Multiple drivers are created when* assignments are made to the same signal in multiple concurrent statements
- Signals with more than one set of projected waveforms must be "resolved"

* multiple drivers are not created by multiple signal assignments in a single process statement

- Example

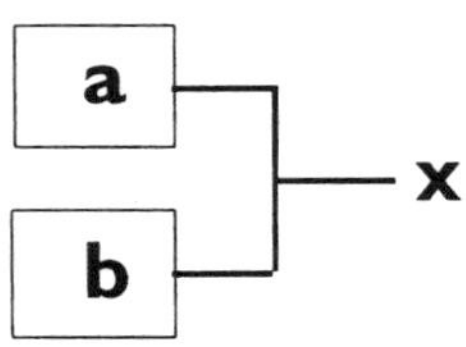


```
architecture ...
begin
    x <= a ;
    x <= b ;
end ;
```

13.3 Multiple Signal Drivers

Signal drivers are created when signal assignment statements are executed. These drivers are containers for the projected output waveforms of the signal. In models created previously in this book waveforms have been assigned to signal objects in only one statement within the model. This was done deliberately.

Consider the following model segment:

```
architecture multiple_drivers . . .
begin
    out_put <= in_put_1 after 20 ns ;
    out_put <= in_put_2 after 10 ns ;
end multiple_drivers ;
```

Suppose *in_put_1* changed to '0' at 0 ns and *in_put_2* changed to '1' at 10 ns. Since it is known that a signal driver is created when a signal assignment statement is executed it is reasonable to assume that two signal drivers are created in this model. One is created when the first signal assignment statement is executed and the other when the second signal assignment statement is executed. Both drivers contain projected waveforms for the signal *out_put*. In fact, both drivers contain a scheduled change for out_put at 20 ns into the simulation. One driver attempts to change *out_put* to '0' and the other attempts to change it to '1'. Obviously, *out_put* cannot assume both values at the same time.

In a sense the output is being "overloaded" with values. Signal objects can have only one current value; yet, in the above example at 20 ns into the simulation of the model, the signal object *out_put* has two conflicting values competing to be the current value. Since it is not possible for both values to take effect there must be some way of choosing one or the other.

Signals with more than one set of projected waveforms must have some way to resolve the competing waveforms and to choose a waveform to take effect. Such signals are called "resolved signals" and have a "resolution" function.

N.B. A signal must be a resolved signal when assignments of waveforms to it are made in two or more concurrent statements. Since a process statement is a single concurrent statement, multiple drivers are not created by multiple assignments to a signal in a single process statement. However, if there are signal assignment statements to the same signal in two different process statements the signal must be a resolved signal.

13.4 Resolution Functions

- Subprograms which return a single value for a signal with multiple drivers
- Invoked whenever signal is active
- Input parameter is one dimensional unconstrained array of signal's type
- Output is single value of signal's type
- Most useful when technology is known

13.4 Resolution Functions

Resolution functions are subprograms which return a single signal value of a signal with multiple drivers. The purpose of these functions is to "resolve" conflicting waveforms for a single signal. When waveform assignments are made to the same signal in two or more concurrent statements multiple drivers are created for that signal. The resolution function of the signal provides the mechanism for determining which waveform will actually be assigned to the signal.

Resolution functions are invoked whenever the signal is active. The input parameter to the functions is a one dimensional unconstrained array whose elements are of the same type as the signal. The output from the function is a single value of the same type as the signal. For example:

type resolve_array **is array** (NATURAL RANGE <>) **of** BIT ;
function resolution_function (in-put : resolve_array) **return** BIT **is** . .

If it was completed, this function could be used as the resolution function for a signal of type BIT. The function is automatically invoked when one or more of the signal's drivers indicate that a change in the value occurs at the current time. When this happens the array in the parameter list is filled with the values from each of the drivers which represents the current value for the signal of that driver. The function must decide what is the actual current value on the basis of the "possible" current values.

The effect of several sources attached to the same wire depends upon the technology of those sources. Unless this technology is known the effect cannot be accurately modelled. In some instances the technology produces a result similar to **or**ing the signals; in other instances the result is similar to **and**ing the signals. For these reasons resolution functions are often given the names "wired_or" or "wired_and". However, resolution functions are not limited to **or**ing and **and**ing values. The effect of multiple drivers can be resolved in any manner the modeller considers best reflects the action of the entity.

13.4.1 Creating Resolution Functions

- **Function name can be whatever seems suitable**
- **Input is one dimensional unconstrained array of same type as "resolved" signal**
- **return type must match signal type**
- **Example**

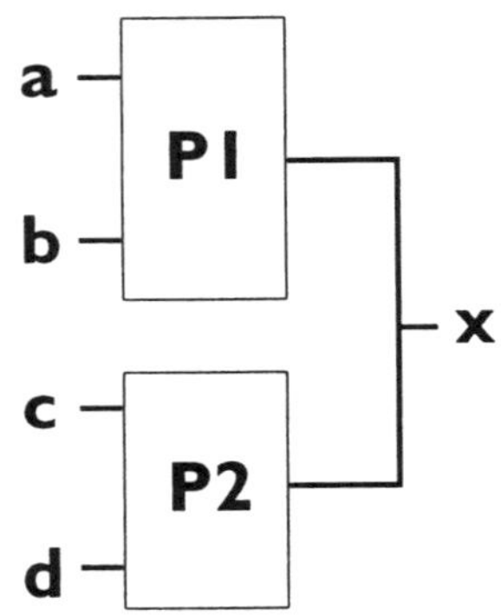


```
function resolve_bits (in_data : BIT_VECTOR)
                                    return BIT is
begin
      for i in in_data'RANGE
            loop
                  if in_data (i) = '0' then return '0' ;
                  end if ;
            end loop ;
      return '1' ;
end resolve_bits ;
```

13.4.1 Creating Resolution Functions

Several important points have been learned about resolution functions:

1. Input to the function is a one dimensional unconstrained array whose elements are of the same type as the type of the signal.
2. The **return** value type must match the signal type.
3. The resolution function is automatically invoked whenever there is a change in the "current" value indicated by one or more of the drivers of the signal.
4. The function name can be whatever the user wishes.
5. Because the technology used in the device affects how "resolution" of multiple sources takes place resolution functions are most useful when the technology being modelled is known.

The way in which the resolution of the competing values is handled depends on the modeller. Consider the following example:

```
function resolve_bits (in_data : BIT_VECTOR)
                                        return BIT is
begin
   for i in in_data'RANGE
           loop
                   if in_data (i) = '0' then return '0' ;
                   end if ;
           end loop ;
   return '1' ;
end resolve_bits ;
```

This resolution function could be used to resolve a signal of type bit which has multiple drivers. It would be invoked when a value in one of the signal drivers matures (a change occurs in a "current" value). The object in the parameter list is of type BIT_VECTOR which is an unconstrained array of type BIT.

To summarise:

1. The "current" values from each of the drivers of the signal are placed in *in_data.*
2. A **for** loop is used to examine each of the values in *in_data.*
3. If any value is '0' the function returns a '0'.
4. Only when all of the possible current values are '1' is a '1' returned

In essence this is a *wired_and* resolution.

13.4.2 Creating Resolved Signals

- **"Resolved" signals have more than one source and have a resolution function**

- **Created during declaration by associating resolution function and type**

- **Example 1**

```
function wired_together (in_data : BIT_VECTOR)
                                   return BIT is . . .
signal res : wired_together BIT ;
```

- **Example 2**

```
function wired (in_data : BIT_VECTOR)
                                   return BIT is . . .
subtype resolved_bit is wired BIT ;
signal example : resolved_bit ;
```

13.4.2 Creating Resolved Signals

A "resolved" signal is one which has more than one source of values and which has a resolution function to "resolve" the values. Simply creating a resolution function does not automatically create a resolved signal. Signals which have multiple drivers must be linked to a resolution function when they are declared. This is done in two ways:

1. The resolution function and type can be linked during the declaration of the signal object. This is done by stating the name of the resolution function in the object declaration. For example:

   function resolve_bits (in_data : BIT_VECTOR)
   return BIT **is . . .**
   signal res : resolve_bits BIT ;

   The resolution function must appear before the signal declaration. "Before" can mean immediately before if the two declarations are made at the same level or it can mean appearance at a higher structural level. Notice that the name of the function comes before the type in the signal declaration and that the type of the signal matches the return type of the function. The function *resolve_bits* could <u>not</u> be used as the resolution function of a signal of type *Std_logic.*

2. The resolution function and type can be linked in a subtype declaration. This is done by stating the name of the resolution function and a type in a subtype. For example:

   function resolve_bits (in_data : BIT_VECTOR)
   return BIT **is . . .**
   subtype inter_res **is** resolve_bits BIT ;
   signal re : inter_res ;

   The subtype declaration links the resolution function and type by preceding the type with the resolution function name. Again, notice that the base type of the subtype must be the same as the return type of the resolution function.

   The function declaration must come "before" the subtype declaration and the subtype declaration must come "before" the signal declaration. When the signal is declared it is declared to be of the appropriate subtype. The advantage of this method is that it saves typing when there are multiple resolved signals of the same type.

The most important aspect of each of these methods is that the base type of the signal must match the return type of the function.

13.4.3 Resolution Function Example

```
entity b_it is
        port (in_a, in_b, in_c, in_d : in BIT ;
                                out_x : out BIT );
end  b_it ;

architecture resolve_example of b_it is
        function resolved (in_data : BIT_VECTOR )
                                    return BIT is
        begin
                if in_data'LENGTH = 0 then return '0' ;
                end if ;
                for i in in_data'RANGE loop
                        if in_data (i) = '0' then return '0' ;
                        end if ;
                end loop ;
                return '1' ;
        end resolved ;
        signal res : resolved BIT ;
begin
        res <= in_a and in_b ;
        res <= in_c and  in_d ;
        out_x <= res ;
end resolve_example ;
```

13.4.3 Resolution Function Example

The model given in Illumination Box 13.4.3 shows how a resolved signal could be set up and used.

The resolution function in this model is basically a "wired_and". The one change between this function and the "wired_and" shown earlier is the addition of a check for a null array. If this function has passed to it an array with no elements it will return a value of '0'. This situation can occur when a disconnect specification has been issued on all sources of the signal.

Because the resolution function is declared below the port clause in the entity declaration, the output port cannot be declared to be a resolved signal. Therefore an intermediate resolved signal is declared within the model and its value assigned to the output port. It is possible to declare a port signal to be a resolved signal, but to do so requires that the resolution function appear before the port clause. This can be done with packages (see Chapter 14).

13.4.4 Overloaded Resolution Functions

VHDL-1993 specifically allows the overloading of resolution functions (See LCS-23, Chapter 3) within the following conditions:

A reference to an overloaded resolution function name in a subtype indication is ambiguous (and is therefore an error) if the name of the function, the number of formal parameters, the result type, and the relationships between the result type and the types of the formal parameters are not sufficient to identify exactly one (overloaded) subprogram specification[1].

Reference

1. *IEEE Standard VHDL Language Reference* Manual (ANSI/IEEE Std 1076-1993).

13.5 Exercises

13.5.1 A certain second order digital filter implements the difference equation:-

$$y_n = c_n x_n + c_1 x_{n-1} + c_2 x_{n-2} - c_y x_{n-2}$$

where the coefficients $c_n + c_1 + c_2 < 1$. The calculated values for the coefficients are:

$c_n = 0.21$
$c_1 = 0.42$
$c_2 = 0.21$
$c_y = 0.1715728$

Assume a 16-bit register limitation and 8-bit analogue to digital and digital to analogue converters with 0-5V range. The input x is sampled at unit times n, n-1, n-2, etc. The output from the ADC and the input to the DAC are unsigned and 8 bits wide. Create a model forthe filter expression y_n.

13.5.2 Create a model which accepts four inputs of Std_logic (a, b, c, d) and which creates one output of Std_logic (e). Inputs a and b are **xor**ed and inputs c and d are **or**ed. The results of the xor and or are tied together for output. Since the outputs are tied into the same signal, there must be a resolved signal. The resolution should take the form of a 'wired_or' condition. A '1' in combination with anything produces a '1'. Two '0's produce a '0'. Any 'X' or 'Z' in combination with 'X, 'Z' or '0' produces an 'X output.

13.5.3 Convert the Std_logic signals into an enumerated type called color [Type color is (black, blue, red, yellow, purple, green, orange)]. 'X' is black, '0' is blue, '1' is red and 'Z' is yellow.Add the converted signals together (use format : new_color <= color_1 + color_2 ;). This will require overloading the '+' operator as follows:

red + yellow = orange
yellow + blue = green
blue + red = purple
black + 'anything' = black
'anything' + 'itself = 'itself'.

13.5.4 Create a model of a device that 'smooths' certain waveforms on a Std_logic line. This device 'watches' the input line for 100 ns and then outputs, on the output line, the waveform which occurred the most times during the 100 ns. The model should record the number of 'X's, '0's, '1's and 'Z's which occur in a 100 ns period. At the end of the sampling period it must output, on another line, the state which occurred the greatest number of times.

Chapter 14

Scope, Visibility, Packages and Libraries

Overview

Declarations are re-introduced in this Chapter and two aspects of declarations which have been touched upon in earlier chapters are introduced formally. These are the aspects of scope and visibility. Scope is the range of the identifying name which appears in the declaration. Visibility refers to where, within its scope, the declaration of an identifier can be "seen".
Packages are a way in which declarations can be made visible across many models.

In this chapter you will learn

- To understand declarations
- The scope and visibility of identifying names
- How to create packages
- How to use user created packages
- How to relate packages to libraries

14.1 Declarations

- **Define meaning of identifiers**

- **Include**
  - **type**
  - **subtype**
  - **object**
  - **file**
  - **subprogram**
  - **component**
  - **entity**
  - **configuration**
  - **package**
  - **alias**
  - **user defined attributes**

14.1 Declarations

Declarations are used to give definition to identifiers. Identifiers are simply names associated with the declaration. Chapter 4 (4.5.1) began by demonstrating how an entity declaration was created. The entity declaration gives a name to a model and defines the relationship between that name and the environment around that name.

Other declarations which have already been discussed include:

- Type declarations
- Subtype declarations
- Object declarations (constant, variable and signal)
- Subprogram declarations (function and procedure)

However, these do not represent all of the possible types of declarations in VHDL. Declarations can also be made in the following areas:

- Files
- Components
- Configurations
- Packages
- Aliases
- User defined attributes

14.1.1 Declarative Regions

- Declarations can be made within specific areas

- Areas
  - design files
  - entity declarations
  - architectures
  - blocks
  - processes
  - configuration specifications
  - subprograms
  - component declarations
  - package declarations and bodies

- Most declarative regions occur after name of area and before begin

- Not all types of declarations can be made in every region

14.1.1 Declarative Regions

Declarations are made within specific areas of the VHDL model structure called declarative regions. These regions exist throughout the VHDL model structure. In fact, the file in which the model is placed is classified as a declarative region. Other declarative regions exist within the following structures:

1. Entity declaration
2. Architecture body
3. Blocks
4. Processes
5. Configuration specifications
6. Subprograms
7. Component declarations
8. Package declarations and package bodies

In many of these structures the declarative region occurs after the name of the region and before the **begin** which appears in the structure. This is true of architecture bodies, blocks, processes, subprograms and entity declarations.

The previous paragraph refers to a **begin** in the entity declaration. Since a **begin** has not appeared in the entity declarations presented in this book, this may cause some confusion. The **begin** and following activity statement area are optional in the entity declaration (See Appendix C, page 476) . In this volume the entity declaration is treated as a declarative region and the optional activity area is not discussed.

Structures such as package declarations and package bodies which do not contain a **begin** are entirely declarative areas.

It is important to realise that not all types of declaration can be in every region. Certain declarations can be made in some declarative regions and not in others.

Signal objects can be declared in the declarative area of the architecture but they may not be declared in a process declarative area.

14.2 Scope and Visibilty

- Where identifiers may occur and how they link to a meaning
- Declarations may have an effect in one part of a model but not in another
- Area of model where a declaration may have an effect is its scope
- Whether or not an identifier corresponds to a particular declaration is its visibility
- Declarations are not necessarily visible throughout their scope

14.2 Scope and Visibility

Two aspects of declarations which have not been formally discussed are scope and visibility. Scope is the range of the name of the identifier within the model and visibility describes to which declaration the name is linked in the model. These two aspects establish where identifiers may occur and how they link to a meaning.

It is important to realise that declarations may have an effect in one part of a model but not in another. Declarations must appear before the identifier in the declaration can be used. A type declaration must appear before the type identifier can be used in an object declaration.

The area of the model where a declaration may have effect is called the scope of the declaration. Since an identifier is simply a name the scope of the identifier is the area where the name can be used.

It is possible to use the same identifier name in multiple declarations. This can be done when:

1. The declaration scopes are parallel.
   Scopes are parallel when the range of one identifier does not overlap the range of the other identifier. An example of parallel scopes would be the declaration of a variable object in one process statement and the declaration of a second variable object in another process statement. Since variable declarations are limited to the process in which they are declared the declarations do not overlap.
2. The declaration scopes are nested.
   Scopes are nested when the range of one identifier is contained within the range of another identifier. An example of nested scopes would be the declaration of a signal object in the architecture and the declaration of a variable object in a process statement. Since the process is within the architecture the scope of the declaration made in the architecture includes the scope of the declaration made in the process.

It is possible that an identifier name cannot be linked to a particular declaration. This is obvious if the identifier appears outside the scope of the declaration. It is not so obvious if the identifier appears inside the scope of the declaration; but it can be explained in that a declaration using the same identifier name as another declaration within which it is nested can hide its outer meaning. This aspect of whether or not an identifier corresponds to a particular declaration is called the visibility of the identifier. Declarations are not necessarily visible throughout their scope.

14.2.1 Scope

- **Range of name of identifier appearing in declaration**
- **Extends from beginning of declaration to end of region in which declaration appears**
- **Limits**

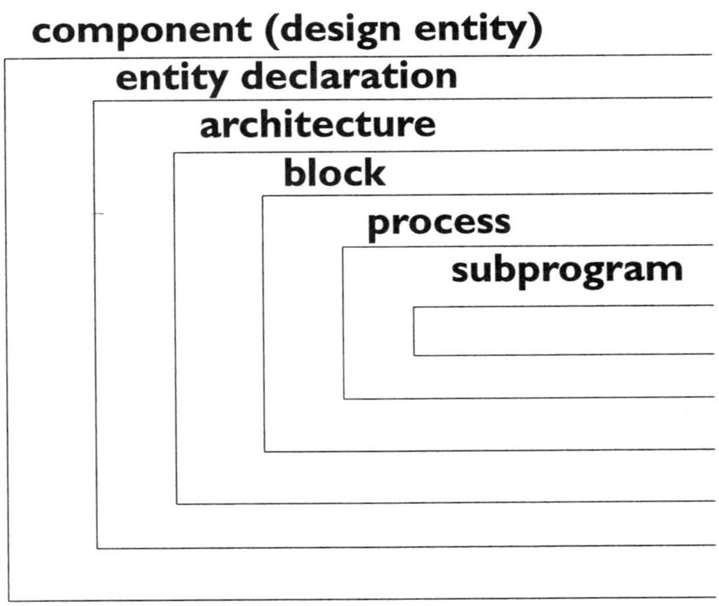

Introduction to VHDL

14.2.1 Scope

Scope has been described as the range of the identifier name appearing in the declaration. It is the region of the model where it may be possible to link an identifier name to the meaning given to the name in a particular declaration. This area extends from the beginning of the declaration to the **end** in the region within which the declaration appears.

Because regions in VHDL occur within other regions, the scope of identifiers tends to follow a hierarchical structure similar to the limits shown in Illumination Box 14.2.1.

1. Declarations made at the design file level extend to all structures within the design file.
   An example is the use of declaration identifiers within packages throughout the structures of the model. The **use** statement links the declarations in the package to the model at the file level. The scope of these declarations extends throughout all structures in the model.
2. Declarations made in the entity declaration extend to all structures in the architectures linked to the entity declaration.
   Although the entity declaration is itself a declaration it can contain other declarations. Declarations made in the entity declaration can be seen by any of the architectures associated with this entity declaration.
3. Declarations made in the architecture extend to all structures in the architecture.
4. Declarations made in blocks within the architecture extend to all structures which exist in these blocks.
5. Declarations made in processes extend to all structures which exist in these processes.
6. Declarations made in subprograms extend to all structures which exist in these subprograms.

The scope of declarations is limited to structures which exist within the structure in which the declaration is made. The scope of the declaration ends at the end of the structure in whose declarative area the declaration is made. Thus the scope of a declaration made in a process ends at the end of the process, but it could extend to a subprogram declaration made in the same process.

14.2.2 Visibility

- Establishes meaning of identifiers
- Declarations are visible only within their scope
- Declarations may be hidden by intervening declarations of identifier

```
ARCHITECTURE
   CONSTANT X : INTEGER := 2 ;
BEGIN
   PROCESS
      VARIABLE X : INTEGER := 3 ;

   BEGIN

   END

END
```

14.2.2 Visibility

Visibility refers to the linking of identifier names with particular declarations. A declaration is said to be "visible" when an identifier name receives its meaning from that declaration. The declaration is "hidden" when the identifier name does not associate with that particular declaration. Two factors concerning visibility must be remembered:

1. Declarations are visible only within their scope.
   An identifier name cannot be linked to a declaration if the identifier is used outside the scope of the declaration. Because a process statement represents a complete scope, a variable object which is declared in one process cannot be seen in another process.
2. Declarations may be hidden by intervening declarations which use the same identifier name.
   Identifier names are always associated with the "nearest" declaration within the structure of the model. In the example in Illumination Box 14.2.2 the identifier name X is associated with the variable object declaration within the process. However, outside the process it is associated with the constant declaration in the architecture. "Nearest" describes the identifier association pattern. An identifier is associated with a declaration in the immediate enclosing structure if a declaration using that identifier name exists. If no declaration exists in the immediate enclosing structure the next structure is checked. This process continues until a declaration is found or until all structures are exhausted and an error results.
   In the example opposite the process is an enclosing structure. Therefore, any reference to X in this structure is associated with the variable declaration. A reference to X which occurred after the end of the process statement would be outside the process structure and would, therefore, be associated with the constant declaration. Within the process the constant declaration would be said to be hidden.

14.3 Mutual Exclusion

- Omitting from consideration all but one declaration for identifier
- Visibility requires determination of unique meaning for identifier

- Two declarations of identifier cannot occur in the same declarative region*
- Multiple declarations of identifier can exist if in nested or parallel declarative regions

*
unless they can be uniquely distinguished, as can enumeration identifiers and subprogram names

14.3 Mutual Exclusion

In the discussion of Illumination Box 14.2.2 the association of declarations is touched upon. This discussion needs further elaboration.

Visibility is tied to the concept of mutual exclusion. When an identifier name is used in multiple declarations the use of the name must be associated with one, and only one, declaration. If more than one declaration could be associated with the name it would not be possible to determine the meaning associated with the use of the name. Visibility requires that all but one declaration of the name of the identifier be omitted from consideration when the name is used.

Because the same identifier name can be used in multiple declarations there must be some way of determining a unique meaning for the identifier.

1. In general an identifier name cannot occur in two or more declarations in the same declarative region.

   Unless there is a way to distinguish uniquely the meaning of an identifier name an error results if multiple declarations using the same identifier name are made in the same declarative region. A unique meaning must exist or the model cannot correctly interrupt the desired action.

   In the case of enumeration identifiers the meaning can be interrupted from the context of the usage. Therefore, the same identifier name can be used in the identifier list of several enumeration type declarations in the same declarative region.

   Subprograms declared at the same level can also be given the same name if they can be distinguished uniquely using the selection rules discussed in Section 13.2.2.

2. Multiple declarations which use the same identifier name can exist if the declarations are in nested or parallel declarative regions.

   A unique meaning can be associated with an identifier name if all but one of the declarations for the identifier can be omitted. If the meaning of an identifier is always the "nearest" declaration then it is possible to have multiple declarations using the same identifier name if the declarations are in nested declarative regions. The "nearest" declaration "hides" the more remote declarations. Declarations in parallel declarative regions are already hidden from each other by scope rules. Hence the same name can be used in declarations which appear in nested or parallel declarative regions.

14.4 Selected Names

- **Declarations which are not visible in a particular region of their scope can still be used**

- **A selected name establishes the location of an identifier's declaration**

- **Format**
  statement_label.identifier_name

- **Example**

```
architecture example of selected_name is
        constant X : INTEGER := 2 ;
begin
        p1 : process (in_1)
                variable X : INTEGER := 3 ;
        begin
                out_1 <= example.X * in_1;
                out_2 <= p1.X * in_1 ;
        end process p1 ;
end example ;
```

14.4 Selected Names

Visibility is very important in that it allows the user to employ the same identifier name in multiple declarations. However, it causes problems when the user wishes to access the value in an object whose declaration is hidden. Consider the following example:

```
architecture example of hidden_names is
    constant X : INTEGER := 2 ;
begin
    p1 :    process (in_1)
                    variable X : INTEGER := 3 ;
            begin
                    out_1 <= X * in_1 ;
                    out_2 <= X * in_1 ;
            end process p1:
end example ;
```

Because the "nearest" declaration of the identifier X in the **process** statement is the variable declaration the value 3 will be multiplied by *in_1* to produce both *out_1* and *out_2*. Suppose that *out_1* should have been multiplied by the constant X and not by the variable X; how could this be accomplished?

Identifier names can be associated with particular declarations by using a construction called a "selected name". Selected names establish the location of the declaration to be associated with a particular identifier name. They consist of the statement label where the declaration occurs, and the identifier name, in the following format:

statement_label . identifier_name

The full stop between the two parts of the selected name is mandatory. In those cases where the statement label has been omitted the default label may be used. In cases where there is no label, such as the architecture declarative area, the name of the structure is used. For example, selected names could be applied to the above architecture as follows:

```
architecture example of hidden_names is
    constant X : INTEGER := 2
begin
    p1 :    process (in_1)
                    variable X : INTEGER : = 3 ;
            begin
                    out_1 <= example.X * in_1 ;
                    out_2 <= X * in_1 ; -- or p1.X * in_1 ;
            end process p1 ;
end example ;
```

14.5 Packages

- Shareable collections of declarations made outside models

- Allow common use of
  - subprograms
  - types
  - subtypes
  - constants
  - signals
  - aliases
  - attributes
  - components
  - disconnection specifications

- Consist of
  - declaration
  - body

14.5 Packages

Packages represent the shareable collections of declarations made outside the model. So far, this text has used declarations made in two different packages in models. Type declarations such as BIT, BOOLEAN and INTEGER are predefined types of the IEEE Std 1076-1993 package STANDARD. Type declarations such as Std_logic) and function declarations to overload logical operators for this type, are made in the package Std_logic_1164.

The purpose of packages is to allow the common use of certain declarations. These declarations include:

- Subprograms
- Types
- Subtypes
- Constants
- Signals
- Aliases
- Attributes
- Components
- Disconnection specifications

By establishing a single declaration for an identifier name which is commonly used across all models mistakes and differences in declarations can be avoided.

Packages consist of two parts:

1. Declaration
2. Body

This is similar to the two parts which constitute subprograms, but there is a notable difference. In subprograms the body is the important part. A body may exist without a declaration. The declaration is needed only in certain instances. The opposite is true of packages. For a package to have any purpose a declaration *must* exist. The body is needed only in certain circumstances.

It should be remembered that packages are created separately from the models. They exist in their own files, they are compiled separately, and they must be linked to the model with the **use** clause.

14.5.1 Package Declaration

- **Describes identifiers declared in package for use outside package**

- **Format**

```
package package_name is
        -- identifier declarations
end package_name
```

- **Allowable identifier declarations**
  - **subprograms**
  - **types**
  - **subtypes**
  - **constants**
  - **signals**
  - **aliases**
  - **attribute specifications**
  - **component specifications**
  - **disconnection specifications**

- **Example**

```
package example is
        type tj is ('T', 'J') ;
        constant pi : REAL := 3.14159 ;
end example ;
```

14.5.1 Package Declaration

The most important part of a package is the package declaration. It describes the identifiers declared in the package for use outside the package. If an identifier is declared in the package body and is not declared in the declaration the identifier cannot be used in a model which accesses the package. A package can be created without a body and still be usable in a model. A body without a declaration is not usable.

The format of the package declaration is:

```
package package_name is
    -- declarations
end  package_name ;
```

The declarations which can appear in the package declaration are:

- Subprogram declarations (but not bodies)
- Type declarations
- Object declarations (signals and constants, but not variables)
- Alias declarations
- Attribute specifications
- Component specifications
- Disconnection specifications
- (**use** clause)*

*The **use** clause is placed in parentheses to indicate that it does not actually appear "within" the package declaration. Instead, it can appear before the package declaration; in the package file. The employment of the **use** clause allows declarations from another package to be referenced in the package. For example, packages which references *Std_logic* must have a ***use*** *IEEE.Std_logic_1164.all;* before the declaration. An example of a package declaration is:

```
package example is
    type tj is( 'T', 'J' ) ;
    constant pi : REAL := 3.14159 ;
end example ;
```

14.5.2 Package Body

- **Describes implementation of subprograms and deferred constants**

- **Format**

```
package body package_name is
        -- declarations, subprogram bodies
end package_name
```

- **Package_name must be same as name used in package declaration**

- **Declarations within body are not visible unless stated in package declaration**

- **Example**

```
package body example is
        function mean (a, b, c : REAL)
                return REAL is
        begin
                return (a + b + c)/3.0 ;
        end mean ;
end example ;
```

14.5.2 Package Body

The package body describes the implementation of subprograms and deferred constants. If subprograms and deferred constants are not declared in the package the package body is unnecessary.

The format of the package body is:

```
package body package_name is
    -- declarations
end package_name ;
```

The package_name must be the same name as that used in the package declaration.

With the exception of subprogram bodies and deferred constants any declarations made in the package body which are not made in the package declaration cannot be used outside the body. Therefore, it is possible to make declarations within the body of the package which cannot be used in models which access the package.

An example of a package declaration and body is:

```
package example is
    function mean (a, b, c : REAL) return REAL ;
end example ;
package body example is
    function mean (a, b, c : REAL) return REAL is
    begin
            return (a + b + c) / 3.0 ;
    end mean ;
end example ;
```

Notice that the name of the package declaration and the body is the same. Notice also that the function declaration appears in the package declaration and that the function body appears in the package body; it cannot appear in the package declaration.

14.5.3 Setting up Packages

- Packages require bodies only if declaring subprograms and/or deferred constants
- Package declarations which contain no identifier declarations are allowed but are of no use
- Package must be compiled to be accessible

14.5.3 Setting up Packages

Three points already made about packages should be emphasised:

1. Packages require bodies only if there are subprograms or deferred constants declared in the package.

   The important part of a package is the declaration. Unless a declaration appears in the package declaration it cannot be accessed by models using the package.

2. A package declaration which contains no identifier declaration is allowed but a package without this declaration would be of no use.

3. Packages must be compiled to be accessible.

   It was stated earlier that a package must be compiled before it can be used by a model. The rule actually goes one step further. The package must be compiled before the model using the package is compiled. If reference is made to an uncompiled package in a model an error will result.

   When compiling packages the package declaration must be compiled before the package body. If the declaration is ever recompiled the body must also be recompiled. However, the body can be recompiled without recompiling the declaration.

14.6 Deferred Constants

- **Appear only in packages**
- **Allow declaration of constant without specification of value**
- **Constant appears without value in package declaration**
- **Full constant declaration must appear in package body**
- **Example**

```
-- package declaration
package example is
    constant constantly : INTEGER ;
end example ;

-- package body
package body example is
    constant constantly : INTEGER := 5 ;
end example ;
```

14.6 Deferred Constants

The discussion of package bodies referred to a VHDL construct called a deferred constant. This construct exists only in package declarations. It allows the declaration of a constant without the specification of a value for that constant.

A deferred constant is created by placing a constant declaration without an explicit value in the package declaration. For example:

```
package example is
    constant constantly : INTEGER ;
end example ;
```

An immediate question which might arise is, "How can one have a constant which doesn't have a value? Surely the whole point of a constant is that it represents a single, unchanging, value. How, then, can something without a value be a constant?". The answer is that the (single, unchanging) value of a deferred constant must appear in a full constant declaration in the **body** of the package. Furthermore, a body must exist for any package declaration which declares a deferred constant. The body for the declaration given above might be:

```
package body example is
    constant constantly : INTEGER := 5 ;
end example ;
```

The next question this raises is, "Why go to all this trouble, why not simply place the value with the constant declaration in the package declaration?". The answer is that to do this would sacrifice one of the powerful features of VHDL, that of flexibility and generality in declaration. Suppose that the constants in a package are used to represent technology parameters for, say, an MOS process. It should be remembered that the package declaration is the important part of the package. All declarations which can be passed to a model must appear in the package declaration. Therefore, the package declarations for the MOS process could be very long and involved. If it was then required that, say, a shrunk version (e.g. 1.5 micron technology down to 1.2 micron technology) be designed, the whole package declaration would have to be rewritten. Using the deferred constant approach all that needs to be changed are be the actual constants, which would then allow the package to represent the new technology. Changing those values requires that the package be recompiled before it is used. If the constants are deferred constants only the package body has to be recompiled. If the constants are STANDARD constants both the package declaration and the package body has to be recompiled. VHDL has what is known as primary and secondary structures. When a primary unit is recompiled all secondary units associated with it must also be recompiled. The recompilation of a secondary unit does not require that its primary unit be compiled. Entity, package, and configuration declarations are primary units; Architecture and package bodies are secondary units.

14.7 Using Packages

- **Identifiers declared in package are not automatically visible to design entities**

- **use makes identifiers in packages visible**

- **Format**
  use package_name.identifier_name ;

- **Example**
  use example.tj, example.mean ;

- **To specify use of all declarations made in a package use .all for identifier_name**

  use example.**all** ;

14.7 Using Packages

The identifiers declared in packages are not automatically visible to design entities (models). Unless the model is told which packages to use and what to employ out of those packages it is not aware of the existence of the package. The **use** statement makes a package, and the identifiers in the package, visible to the model. The format of this statement is:

use package_name . identifier_name ;

This form of the **use** statement identifies the name of the package which is accessed by the model and the name of the identifier within the package which is used in the model. An example of **use** statements which have been included in models written in this book looks like this:

use IEEE.Std_logic_1164.**all** ;

The mgc_portable part of the statement indicates the location of the package. If a location is not specified the model assumes that the package is in the same directory as the model itself. The **all** indicates that all of the identifiers in the package are to be visible in the model. Consider the following examples:

use example . tj ;

This **use** statement indicates that the model has access to the declaration for the identifier called *tj* which is located in the package example. Because no location is given the package must be located in the same directory as the model;

use example . tj , example . mean ;

This **use** statement indicates that the model has access to the declaration for both the identifier called *tj* and the identifier mean which are located in the package example.

An advantage of being able to pick specific identifiers rather than all the identifiers is related to the question of visibility. If two packages contain different declarations of the same identifier name, one will hide the other if .**all** is used. While this might be acceptable in some cases there are instances when one, or the other, only must be picked. By specifically omitting the name of the identifier in one **use** clause and including it in the other the question of visibility is not an issue.

Another advantage of being able to pick specific identifiers rather than all of the identifiers is related to how material from packages is included in the model. It should be remembered that packages must be compiled before they can be used. This is because a package used in a model is actually copied into the compiled model code. This has some interesting effects. Firstly, if .**all** was used the whole package is copied not just the compiled model. This can make for a large compiled model even though only one or two items may actually be used from the package. Secondly, if a package is changed, as long as the model is not recompiled, the model is not automatically updated with the contents of the "new" package.

14.8 Predefined Packages

- IEEE Std 1076-1993 packages
  - STANDARD
  - TEXTIO
- IEEE
  - Std_logic_1164 (not in LRM)
- Vendor supplied packages (examples)
  - Math
  - qsim_logic
  - qsim_relations

14.8 Predefined Packages

Certain packages in VHDL are described as "predefined". Only the first two listed below are to be found in the IEEE Std Language Reference Manual (IEEE Std 1076-1993.

- STANDARD
  The package STANDARD is defined in the IEEE Standard VHDL Language Reference Manual (IEEE Std 1076-1993). It contains the predefined types mentioned in Section 5.3.3, the subtypes NATURAL and POSITIVE, and the function **now**. An implicit context clause naming this package is assumed to exist at the beginning of each design unit. Package STANDARD may not be modified by the user.
- TEXTIO
  The contents of TEXTIO is defined in the IEEE VHDL LRM. This package contains declarations of types and subprograms that support ASCII I/O operations.
- IEEE.Std_logic_1164
  This package supports a "standard" 9 state type definition for model simulation that all modellers may use. By using a standard type, models can be passed easily from simulator to simulator. It is not part of the LRM.
- MATH
  This package contains many common mathematical functions. For example, a square root function may be found in this package.
- Qsim_logic
  This is a MGC supplied package. It provides the declarations commonly needed by the MGC design environment.
- Qsim_relations
  There are questions about how certain aspects of modelling should be done. For example, should the comparison 'X' = 'X' return TRUE or FALSE? In some models the answer might be TRUE but in others FALSE. In any case since 'X' is unknown it cannot be said with any certainty that the value of the two 'X's should be the same. In *qsim_logic* the package overloads the = such that TRUE is returned. However, *qsim_relations* overloads = such that FALSE is returned. The choice is therefore given to the user by means of the selection of the relevant package.

14.9 Libraries

- Provide storage for compiled design units

- Units
  - primary
    - entity declarations
    - package declarations
    - configuration declarations

  - secondary
    - package bodies
    - architecture bodies

- Classifications
  - working library
  - resource library

14.9 Libraries

Libraries provide storage for compiled design units. As stated earlier, there are two classes of design units.

1. Primary Units
   - entity declarations
   - package declarations
   - configuration specifications
2. Secondary Units
   - architecture bodies
   - package bodies

A secondary unit must be placed in the same library as the primary unit it references. Recompilation of a primary unit requires recompilation of the secondary units referenced to it.

There are two general classes of library:

1. The Working library

   The working library corresponds to the working directory. It is always the directory the system is currently accessing. The name given to the working library is WORK.
2. Resource Libraries

   Resource libraries offer additional storage for design units. Any local name may be given to resource libraries

The statement in Illumination Box 14.9 declares that libraries provide storage for compiled design units. While this is true, it should also be noted that source code files can also be stored in libraries. Since the mixed storage of source code and compiled code can become "untidy" source code is often stored separately from compiled code when "resource" libraries are created.

14.9.1 Working Library

- **Corresponds to working directory**
- **Is named "work"**
- **Units in working library are directly accessible**
- **Identifiers within packages in working library are made visible with use**

  use example.**all** ;

  or

  use work.example.**all** ;

14.9.2 Resource Libraries

- **Offer additional storage for compiled design units**
- **Name is logical name and must be mapped to physical structure**
- **Units in resource libraries are not directly accessible**

14.9.1 Working Library

The working Library corresponds to the working directory and has the logical name WORK.

Because compiled units in the working library are directly accessible there does not have to be a LIBRARY statement declaring the logical name of the working library. Identifiers within packages in the working library are made visible with the **use** statement. The name WORK does not have to appear in the **use** statement but may be placed there for documentation purposes, e.g.:

use example . **all** ;

or

use WORK.example . **all** ;

Each of the examples make all the declarations made in a package named *example* visible in the model. This package must be in the working library (i.e. the working directory).

14.9.2 Resource Libraries

Resource libraries offer additional storage for compiled design units. Except for the library in which the VHDL package STANDARD is stored primary and secondary units in these libraries are not directly accessible. Two things must be done to make the items in these libraries visible to a model:

1. The library must be given a logical name within the model.
   This is done using the **library** statement to be discussed overleaf.
2. The logical name must be mapped to a physical structure.
   This is done during compilation of the model and is discussed in Section 14.9.4.

Resource libraries correspond to directories. Any directory that is not the working directory can, potentially, be a resource library.

Notice that one of the items in Illumination Box 14.9.2 refers to "compiled design units". Directories, and thus libraries, can contain both compiled and uncompiled units. However, only compiled units can be referenced from within another unit. Therefore, for units within a library to be accessible they must be compiled.

14.9.3 Accessing Libraries

- **library statement identifies logical name of library to be accessed**

- **use statement makes design units within library accessible**

- **Format**

```
library library_name ;
use library_name.package_name ;
```

- **Identifiers within packages within a library are accessed with a second use statement**

- **Example**

```
    library things ;
    use things.func_pkg ;
    use func_pkg.all ;
-- alternative
    library things ;
    use things.func_pkg.all ;
```

14.9.3 Accessing Libraries

Except for the library WORK, and the library STD in which the VHDL package STANDARD is stored, libraries must be given a logical name within the unit accessing the library. This name identifies the library within the accessing unit. The name must follow the standard VHDL naming rules but may be anything the modeller chooses.

Logical names are assigned to libraries with the **library** statement:

library logical_library_name ;

This statement must precede any use of the logical name of the library.

A library can contain several different packages and the use of a library statement does not make a particular package in that library accessible. Packages within a library are made visible to the accessing unit by means of a **use** statement.

use logical_library_name . package_name ;

In addition, the modeller must state what identifiers within a package are to be used in the model. This can require a second **use** statement.

use package_name . identifier_name ;

Thus the following would be required if a modeller wished to use all of the identifiers, in a package called *func_pkg*, in a library which the modeller has given the logical name *things*:

library things ;
use things . func_pkg ;
use func_pkg . **all** ;

Because the two **use** statements can be combined this can also be written as:

library things ;
use things . func_pkg .**all** ;

A library statement does not have to appear in a model when both the compiled version of the model and the compiled version of the package exist in the working directory. In addition, neither a library statement nor a **use** statement have to appear in a model using the VHDL package STD. This is because the following implicit statements are assumed to exist at the beginning of every design unit:

library WORK ;
library STD ;
use STD . STANDARD . **all** ;

14.9.4 Package/Library Example

```
package func_pkg is
    function mean (a, b, c : REAL)
                    return REAL ;
end func_pkg ;

package body func_pkg is
    function mean (a, b, c : REAL)
                    return REAL is
    begin
        return (a + b + c)/3.0 ;
    end mean ;
end func_pkg ;
-----------------------------------------------------
library things ;
use things.func_pkg.all ;
entity average is
port (in_1, in_2, in_3 : REAL ;
              signal_out : REAL ) ;
end average ;
architecture pkg_example of average is
begin
    signal_out <= mean (in_1, in_2, in_3) ;
end pkg_example ;
```

14.9.4 Package/Library Example

The model in Illumination Box 14.9.4 shows an example of a user created package and a user created model which uses that package.

- This device accepts three input signals whose values are real numbers. It finds the average of the three numbers and outputs the result.
- A function which returns the average of three real numbers is contained in the package *func_pkg*.
- The first two statements in the model identify a logical library called *things* and state that all the declarations in a package called *func_pkg* within things are to be accessible by this model.
- The model contains one concurrent statement. This statement calls the function mean and assigns the return value to the output signal.
- The package *func_pkg* has both a package declaration and a package body. It must have both a declaration and a body because a function is declared within the package. The function declaration must appear in the package declaration so that the function will be visible to the unit accessing the package. The body of the function must appear in the body of the package according to VHDL structural rules. Hence the package has both a declaration and a body.
- The package must be compiled before the model is compiled.
  It should be remembered that the source file of a model must be placed in a directory with the same name as the entity declaration name used in the model. Similarly, the package source file must be placed in a directory with the same name as the package name. The directory containing the package directories is the physical structure which maps into the logical library name.

14.10 Exercises

14.10.1 Create a package called chap_14_pkg and within it declare a type called "three_state", a function that converts "three_state" signals to Std_logic signals, and a function that overloads the **and** operator to compare three_state signals.

(a) The type three_state consists of 'U', 'L' AND 'H'.

(b) Std_logic signals map to three_state signals in the following manner:

three state	Std logic
U	X
L	0
H	1

(c) When overloading the **and** operator, use the following table as a guide to mapping the relationships:

R\L	U	L	H
U	U	L	U
L	L	L	L
H	U	L	H

14.10.2 Create a package called UP_EEE which overloads the IEEE STANDARD types BIT and BIT_VECTOR, for the predefined operators shown below, with the values (0, 1, X, Z). Use the simple_name *four_val* for the values. Add functions for

i) the simple connection of two drivers
ii) the simple wire connection of multiple drivers
iii) wired_AND for multiple drivers
iv) wired_OR for multiple drivers

to the set.
Predefined operators : "and", "or", "nand", "nor", "xor", "xnor", "not".

14.10.3 Add to the package of Ex. 14.10.2 the functions and procedures created in exercises 11.5.6, 12.7.3 and 12.7.4.

14.10.4 Add to the package of Ex. 14.10.2 a function which will increment a Std_logic_vector. Hint: Use the functions created in Ex. 11.5.6.

Chapter 15

TEXTIO, Blocks and Guards

Overview

Textual input and output refer to the process of reading from, and writing to, text files. The package TEXTIO contains declarations of types and subprograms that support formatted ASCII I/O operations. A block represents a portion of the hierarchy of a design. A block is either an external block or an internal block. A guard expression is an expression associated with a block statement which controls assignment to guarded signals within the block

In this chapter you will learn

- How to identify, read/write, and process files
- How to partition models using blocks
- How to create guarded assignments

15.1 Textual Input and Output

- **Provides a method for reading from and writing to text files**
- **Text files are ASCII files**
- **ASCII files**
  - **consist of lines of characters**
  - **lines end with a carriage return**
- **TEXTIO package**
  - **contents specified by the IEEE**
  - **declares types and subprograms for file manipulation**

15.1 Textual Input and Output

Textual input and output refers to the process of reading from, and writing to, text files. The text files involved are ASCII character files.

ASCII character files

1. Consist of lines of ASCII characters
2. Each line of characters ends with a carriage return
3. Individual groups of characters (data items) on the line are separated from the next group of characters by a blank character (space)

The processing of text files is facilitated by the package TEXTIO:

1. The contents of TEXTIO is specified as part of IEEE Std 1076-1987
   This package is not meant to handle all file processing needs. It contains only the functionality needed to process simple textual input and output associated with the types declared in the IEEE package STANDARD.
2. The package TEXTIO consists of the type and subprogram declarations needed to process simple text files.
   Although the contents of the package TEXTIO are mandated by the IEEE this package is not automatically included in a model. To make the declarations in this package visible in a model a **use** statement specifying the package TEXTIO must be placed in the model.

15.2 Package TEXTIO

- **IEEE predefined package providing textual input and output capabilities**

- **Type declarations for text I/O**

  - **type** line **is access** string ;
  - **type** text **is file of** string ;

- **Procedures for reading from files/lines**

  - **procedure** READLINE.....
  - **procedure** READ.....
    bits, bit_vectors, boolean, character, integer, real, string, time

- **Procedures for writing to files/lines**

  - **procedure** WRITELINE.....
  - **procedure** WRITE.....
    bits, bit_vectors, boolean, character, integer, real, string, time

•**Function for locating end of file**

  - **function** endfile (op : **in** LINE)
    return BOOLEAN ;

15.2 Package TEXTIO

The IEEE specifies the contents of a package called TEXTIO for textual input and output capabilities. Unlike the IEEE package STANDARD it is not automatically included in every model. It must be included in a model by means of the **use** statement.

Several type declarations and procedure declarations which facilitate the processing of textual data are made in TEXTIO. These include:

1. Type declarations
   type line and **type** text
2. Procedures for reading from and writing to files
   READLINE and WRITELINE
3. Procedures for reading from and writing to lines
   READ and WRITE
4. Functions for locating the ends of files
   ENDFILE

The above items are introduced and discussed in the following pages.

The reason that TEXTIO is not automatically included in models is in order to avoid unnecessary code which does not manipulate files.

To include this package in a model it must be placed at the beginning of the model by means of the **use** statement:

use STD .TEXTIO . **all** ;

Notice that it is not necessary to include a **library** statement to identify the library STD.

15.3 Identifying Files

- **Files are declared in declarative regions**
- **File declaration identifies**
  - **logical name of the file**
  - **type of file**
  - **mode of file (in or out)**
  - **physical file by path name**
- **Format**

  file logical_file_name : **type is** mode
  "/physical/path/file_name" ;

- **Example**

  file file_name : TEXT **is in**
  "/user/training/foo_file" ;

15.3 Identifying Files

If a file is to be used within a model it must be identified with that model. This is done using the file declaration statement:

file file_name : TEXT **is in** ". ./path_to/physical_file" ;

This statement supplies the model with several important items concerning the file:

1. The logical name of the file

   The logical name of the file is the name following **file** in the statement. Since this is a logical name it can be anything suitable for the model. This is the name that is used by READLINE or WRITELINE statements.

2. Type of file

   The type of file identifies the manner in which data has been placed in the file. The package TEXTIO specifies only one type of file: TEXT. Text files are files of strings; thus text files have data transferred to them by groups of characters called strings. For the sake of simplicity a string is equivalent to a physical line in this book. Other types of file are possible but are not discussed here.

3. Mode of file

   Only two modes are permitted: **in** and **out**. A file opened as mode **in** can be read from but not written to, and a file opened as mode **out** can be written to but not read from; therefore, since files can only be opened with one mode, a file containing both mode **in** and **out** cannot be opened.

   This last rule produces some interesting restrictions on processing text files. As soon as a file is closed the pointer is moved back to the beginning of the file. Therefore, when a file is opened as mode **out,** data written to the file is written at the beginning of the file. This clears any data already in the file and so data may not be appended to a file (unless this is a definite intent of the process).

4. The physical file associated with the logical name

   It has been learned that the physical library is linked to the logical library name in the **library** statement during compile by setting the compiler option. This is not the case with files. The physical file associated with the logical file name is identified in the **file** statement. If the file does not exist in the location specified the system will create the file. If a file is written to and it cannot be found after execution of the model it may be that the path to the **file** statement specified a location other then the one being looked at.

15.3.1 Reading/Writing Files

- Information is transferred to and from files line by line

- READLINE reads a line from a file
- WRITELINE writes a line to a file

- Format

- READLINE (logical_file_name, line_name);
- WRITELINE (logical_file_name, line_name);

- Data in the model is read from or written to fields in the line

- READ reads a value from a line
- WRITE writes a value to a line

- Format

- READ (line_name, object_name) ;
- WRITE (line_name, object_name, [justification, field_width]) ;

15.3.1 Reading/Writing Files

Information is transferred to and from files line by line. The key point here being that data in files is read from the file by a physical block called a line. In an ASCII file a physical line is any number of characters from the beginning of the line to the "return" character at the end of the line. Only complete physical lines can be transferred from files.

The transfer of lines of data to and from files is done with the READLINE and WRITELINE statements. In VHDL terms these are not actually statements. READLINE and WRITELINE are **procedures** within the package TEXTIO. Since the only procedures for reading and writing files discussed in this book are those in TEXTIO the limitations on file manipulation are those imposed by TEXTIO. Thus the reading and writing of files is done by a procedure call and procedures are called from a statement.

READLINE is used to transfer a physical line in the file to an object of type LINE in the model. It has no interest in the content of the line and involves only a transfer of ASCII characters. Therefore, to read a line from a file, an object of type LINE must be declared in the model.

WRITELINE is similar to READLINE in that it transfers a line of characters to a file.

Individual values cannot be written to, or read from, a file; only lines of characters can be transferred.

The data in an object of type LINE can be read one item at a time with the READ procedure in TEXTIO. This procedure locates and moves a value from the data line in sequential order. The READ procedure requires that each item of data be separated by a blank space. This identifies a single data item to it. To reach the third data item on the line the READ procedure must be executed three times. There is no method of specifying a particular data item. READ also has an additional function; it must "convert" the characters that it reads to the type of the object to which the value is being assigned. Therefore READ is overloaded to "read" various types. For example, one READ is needed for type BIT and another for type BOOLEAN. Because TEXTIO is mandated by the IEEE the only types it recognises are those "predefined" in the STANDARD package. Therefore, READ is overloaded only for the types BIT, BOOLEAN etc. READ cannot be used to read a type such as "colour", which is user declared. READ could, however, be overloaded to perform this action.

15.3.2 Processing Files

- Include TEXTIO in model
  use STD.TEXTIO.all ;
- File is opened when the area it is declared in is elaborated, it is closed when the area is exited
- A file can be opened for read or write but not for both at the same time
- A line from a file can contain one or more values
- Each value in a line must be read before a new line is read
- The READ procedure from TEXTIO is overloaded for different types
- Fill a line with values before writing the line to a file

15.3.2 Processing Files

The following are important factors to be considered when TEXTIO is used in models to process files:

1. The package TEXTIO is not automatically included in a model. A USE statement must be placed in the file to access TEXTIO. The form of the statement is:

   use STD.TEXTIO.**all** ;

   Note that STD is the library name of the location of TEXTIO. Since every VHDL system must know where STD is located (an IEEE requirement; STD is a reserved identifier) no library statement is needed to identify it nor is it necessary to set a compiler option to point to STD on compile.
2. Files are opened when the **file** statement is encountered in a particular declarative area and they are closed when the limit (or end) of the scope of the declaration is exited.
   This is important because it means that a file can be opened and closed many times during the execution of a model. Suppose that the **file** statement appeared in a subprogram. The file would be opened every time the subprogram was called and it would be closed every time the subprogram was exited. This might mean a large number of openings and closings during the course of model simulation.
3. Files cannot be opened for input and output at the same time.
   Files can have a mode of **in** or **out** but not **inout**. This will severely limit some applications of files. For example, one of the primary limitations will be that files cannot be appended (i.e. data cannot be added to the end of data already in the file).
4. Lines in files can contain many data items.
   Data is read from and written to files line by line. A line is a physical line and there is no limitation on the number of data items on a line. However, if a data file is created using an ASCII editor, each data item on a line must be separated by a blank space and the end of the line must be a carriage return.
5. Data is moved from the file to the model using the READLINE statement. Once data is moved from the file to an object of type LINE in the model it must be read into an object of the type of the data using the READ statement. The data on a line is erased as soon as READLINE is used to read a new line from a file. Therefore, all data must be read from a line before a new line is read. The same is true of writing a line to a file. As soon as WRITELINE is used the line that was written from is erased so that new data may be entered on it.

15.3.3 File Example

```
use STD.TEXTIO.all ;
entity example is
        port ( control : in BIT ) ;
end example ;

architecture behav of example is
begin
        process
                file data_in : TEXT is in
                        ". . /input_data_file" ;
                file data_out : TEXT is out
                        ". ./output_data_file" ;
                variable in_line, out_line : LINE ;
                variable inverse : BIT ;
                variable half : INTEGER ;
        begin
                wait on control ;
                while not (ENDFILE (data_in))
                        loop
                                READLINE (data_in, in_line) ;
                                        READ (in_line, inverse) ;
                                        READ (in_line, half) ;
                                                inverse := not inverse ;
                                                half := half/2 ;
                                        WRITE (out_line, inverse) ;
                                        WRITE (out_line, half) ;
                                WRITELINE (data_out, out_line) ;
                        end loop ;
        end process ;
end behav ;
```

15.3.3 File Example

The example shown in Illumination Box 15.3.3 presents the basics on using the declarations in TEXTIO to read from and write to ASCII text files.

Note that the IEEE package TEXTIO is included in the model:

use STD.TEXTIO.**all** ;

Since TEXTIO is in the library STD, a **library** statement does not have to be included in the model.

The **file** statements which appear in the **process** statement identify the files that are to be used in the model. When these statements are first executed the system looks for the files and, once found, they are opened for either input or output depending on the mode identified in the statement. If the files do not exist the system creates files with the names given in the location identifier of the **file** statement. In this case the file names would be input_data_file and output_data_file.

Data is read from files and written to files line by line. Therefore, it is necessary to declare two objects of type LINE to read and write data. This is done in the line:

variable in_line, out_line : LINE ;

The type LINE is declared in the package TEXTIO.

The actual reading of the data from the input file is done within the **while** loop. This loop continues to process data until the end-of-file mark is encountered in the input file.

READLINE reads a line of data from the input file. The procedure READLINE is declared in the package TEXTIO. It should be remembered that data is transferred into and out of files by physical lines in the file.

Data is read from the line by the READ procedure. This procedure is also declared in the TEXTIO package. The procedure is overloaded to "read" data of many different types. The data being read in this case is of type BIT and type INTEGER. READ is overloaded for both of these types.

WRITE places data into an object of type LINE. WRITE cannot be used to write data directly to a file; only complete lines may be written to files. Write is overloaded in the same way as READ.

WRITELINE writes lines to files.

When the **end process** is encountered the files are closed. Thus, if the process were to be re-executed, the results would be exactly the same. The input file would be re-opened and read from the beginning. The output file would be re-opened and written to from the beginning. Thus the results in the output would not contain two copies of the output; it would contain only one set of output.

15.4 Design Partition

- **Dividing model into smaller parts reduces complexity to manageable levels**
- **Methods**
  - **entities**
  - **subprograms**
  - **blocks**
  - **processes**

15.4 Design Partition

It is in the nature of things that tasks which appear, on the surface, to be complex or lengthy tend to be avoided. The tasks which are completed tend to be those thought to be "manageable".

Design partition is the process of dividing a model into smaller parts. This process should aim to reduce the apparent complexity to what the modeller can accept as "manageable" levels. In addition, it can, if used wisely, make the model easier to develop and maintain. For example, a machine vision system is a complex electronic system for capturing video images, processing those images, and creating outputs based on the results of the analysis of the images. Such a system may consist of a series of connected boards of which each has a specific function. It is possible that the function of an individual board can be subdivided into subfunctions and that those subfunctions can be further reduced with reference to the type of ICs needed to accomplish the subfunctionality. It may even be possible to break the function of the ICs into their subfunctions and so on. At some point in this process the functionality of the machine vision system will have been resolved into "familiar", or "manageable", levels.

Modelling can be performed in a similar manner to the above. The process of creating a system design model may start with a very high level conceptual model which is then broken down into behavioural and then to functional parts as detail and structure is added to the model (top-down approach). On the other hand an existing, or known, system process might be made simpler by creating the functional structures before beginning the modelling process (bottom-up approach).

The one_bit_adder model in Section 9.9 shows functional partition. There are the equivalent of two process statements in this model. The first determines the number of '1's occurring on the three inputs and the second assigns the correct output based on the number of '1's. In this case process statements are used to accomplish the partition (the selected signal assignment is a concurrent statement. Concurrent signal assignments are the equivalent of a sequential assignment within a process statement).

Subprograms are another method of breaking large segments of code into smaller segments. They represent units of functionality. The creation of subprograms is most important when the same unit of functionality is a unit in several different locations in the model. Even the behavioural design entity (model) itself represents a unit of functionality. Such units can be linked together in structural models and can be repeated many times within such models.

In VHDL there is a fourth structure which serves the dual purpose of providing a method of design partition and a method of execution control. This structure is called a block and is the subject of discussion of the rest of this chapter.

15.5 Block Statement

- **Allows grouping of concurrent statements into logical units ***

- **Blocks can be nested to further support partitioning of model**

- **Format**

  - block_label : **block** [(guard_expression)][**is**]
    -- declarative_statements
    begin
    -- concurrent statements
    end block [block_label] ;

•Example

```
block_example : block begin
                    out_b <= in_1 or in_2 ;
                end block block_example ;
```

* **logical units describe a portion of model**

15.5 Block Statement

The block statement provides a method of creating groupings of concurrent statements and control signal assignments. Statement groups can be logical units which represent a portion of the functionality of the model. In this way blocks can be used to support the partition of a design into functional units.

Blocks may also be used to control the execution of concurrent statements. An optional aspect of blocks is the creation of a boolean signal based on an expression in the block statement. This expression is known as a guard_expression. The **guarded** clause can be used to examine the boolean signal based on the guard expression and to determine whether or not the statement is to be executed.

In some ways blocks are the concurrent equivalent of process statements. Blocks enclose a functional grouping of concurrent statements. Processes enclose a functional grouping of sequential statements. However, because **block** statements are themselves concurrent statements, blocks can be nested within blocks to further support the partitioning of the model. Since **process** statements are concurrent statements, and since only sequential statements can be placed within process statements, process statements may <u>not</u> be nested within other process statements (it should be noted that **process** statements can appear with **block** statements since any concurrent statement may appear in a block).

The format of the block statement is:

```
block_label :   block [(guard_expression)]
                    -- header_statements
                    -- declarative_statements
                begin
                    -- concurrent_statements
                end block [block_label] ;
```

The label that appears on the block must be supplied by the modeller; there is no default label for blocks.

The guard_expression is optional. Because a boolean type signal is created by the guard expression it must evaluate to a boolean TRUE or FALSE. This signal is called GUARD and is discussed later in this chapter.

The header statements in a block can be used to support an "environment" interface representation. Both port and generic statements may appear here.

Any of the declarative statements in a block discussed in this book may appear between **block** and **begin**. The general rules of scope and visibility apply to declarations made in this area. For example, an object declared in this area cannot be seen outside this block but may be seen within levels in the block such as **process** statements.

15.5.1 Using Blocks

- **Blocks can simply label sections**
- **Example**

```
•
•
•
d_q <= in_1 after 5ns
block_1 :
block begin
        not_d_q <= not d_q after 1 ns ;
end block block_1 ;
•
•
•
-- is exactly the same as :

•
•
•
d_q <= in_1 after 5 ns ;
not_d_q <= not d_q after 1 ns ;
```

15.5.1 Using Blocks

Blocks have two uses:

1. To partition a design into logical units and add "structure" to a model.
2. To provide execution control of concurrent signal assignment statements.

In its simplest form a block merely adds a label to a section of a model. For example, consider the following two code segments:

```
d_q <= in_1 after 5 ns ;
block_1 :
    block begin
            not_d_q <= not d_q after  1 ns ;
    end block block_1 ;
```

and

```
d_q <= in_1 after 5 ns ;
not_d_q <= not  d_q after 1 ns ;
```

The function of the two code segments is identical. The block version simply supplies a label to the second statement of the second version.

In the above case a block does not actually need to be used to label the second concurrent signal assignment statement as these can carry their own labels. However, any number of concurrent statements could appear between **block begin** and **end block**.

Blocks can do more than simply label code segments. Consider the following example:

```
block_2 : block
                signal in_blck : Std_logic := 'X' ;
          begin
                in_blck <= input_1 after 1 ns ;
          end block block_2 ;
block_3 : block
                signal  in_blck : Std_logic := 'X' ;
          begin
                in_blck <= input_2 after 1 ns ;
          end block block_3 ;
```

The signal *in_blck* that appears in block_2 is a different signal from that of *in_blck* in block_3. In this case the blocks provide a method of limiting the scope of a signal to a specific area of the model.

15.5.2 Example of Partitioning

```
architecture parts of microprocessor is
begin
accumulator : block begin
                    -- concurrent statements
              end block accumulator ;
program_counter : process
                        -- sequential statements
                 end process program_counter;
instruction_register: block begin
                        -- concurrent statements
              end block instruction_register ;
instruction_decoder : block begin
                        -- concurrent statements
              end block  instruction_decoder ;
stack_pointer : block begin
                    -- concurrent statements
              end block stack_pointer ;
                    •
                    •
                    •
end parts ;
```

15.5.2 Example of Partitioning

block and **process** statements can be used to create logical partitions of models.

The example in Illumination Box 15.5.2 shows how a model of a microprocessor might be partitioned by functionality. Notice that in this example the functions of the accumulator, instruction_register, instruction_decoder, and stack_pointer are expressed as concurrent statements within blocks. The function of the program_counter is expressed in a series of sequential statements within a process.

Whether **block** or **process** statements are used to create the partition of the design depends upon the complexity of the segments and/or the preferences of the modeller. In some cases a simple process is sufficient to model the functionality; in others several processes may have to be grouped within a block.

The use of blocks mandates the use of labels. Blocks cannot be created without user-supplied labels. These labels should be chosen such that they reflect the functionality of the block. The labels on the facing page are examples of "meaningful" labels.

If **process** statements are used in conjunction with blocks to partition the design they should also be labelled. Even though a user-supplied label is not required with **process** statements, a meaningful label should be used. A **process** statement without a label which appeared among labelled block statements might be confusing as its purpose would be unknown and might be difficult to discern.

15.5.3 Guard Expression

- **Appears in parentheses after block in block statement**
- **Format**
  block_label : **block** (guard_expression)
- **Creates signal of type boolean named GUARD**
- **GUARD is used like any other signal but may not be explicitly changed**
- **Example**

```
guard_example : block (clk = '1')
        begin
            -- concurrent statements
        end block guard_example ;
```

15.5.3 Guard Expression

A block statement may be written with a guard expression. This expression appears in parentheses after **block** in the block statement. It creates a signal of type boolean, named GUARD:

```
block_label : block (guard_expression)
```

The guard_expression can be any expression that produces a boolean value (TRUE or FALSE). Since the result of relational comparisons is a boolean value, guard_expressions are often written using relational operators. For example:

```
guard_example : block (clk = '1')
        begin
                -- concurrent statements
        end block guard_example ;
```

Any other expression that produces a boolean result could also be used. For example, the logical comparison of two boolean objects produces a boolean result and could be used:

```
guard_example : block (boolean_1 or boolean_2)
```

Since a block statement is a concurrent statement the guard expression must be constructed using signals or constants. If constants are used the value of GUARD never changes. If signals are used the value of GUARD is recalculated any time one or more of the signals in the guard_expression changes.

Although the GUARD expression looks somewhat like the sensitivity list in a **process** statement it does not function in the same manner. The GUARD expression does not control whether or not the block is executed. Blocks contain concurrent statements and these activate at the same time whether or not they are in a block. The guard expression is not a sensitivity list which controls execution.

The signal GUARD may be used like any other boolean signal but it may not explicitly be changed. For example, it may be used in an **if** statement but it may not have a value assigned to it:

```
if GUARD THEN          -- is a valid use of GUARD
    GUARD <= FALSE -- is an invalid use of GUARD
end if ;
```

Since GUARD is a signal any of the signal attributes may be applied to it and it may be used in any statement where signals are commonly used. For example:

```
a_process :     process
                begin
                        in_block <= input_1 after 1 ns ;
                        wait on  GUARD ;
                end process a_process ;
```

15.5.4 Guarded Assignment

- **Used to control assignment of signal values**
  - **conditional signal assignments**
  - **selected signal assignments**

- **Appears as reserved word guarded immediately to the right of the assignment operator**

- **Assignment is made when**
  - **GUARD changes to TRUE**
  - **signal changes and GUARD is TRUE**

- **Example**

```
guard_example : block (clk = '1')
begin
    signal_2 <= guarded signal_1
end block guarded_example ;
```

is equivalent to

```
equivalent : process
        begin
            if GUARD then
                signal_2 <= signal_1 ;
            end if ;
            wait on GUARD, signal_1 ;
        end process equivalent ;
```

15.5.4 Guarded Assignment

The option **guarded** is used in conjunction with GUARD signals to control the assignment of signals in concurrent signal assignment statements such as conditional signal assignment statements and selected signal assignment statements.

guarded cannot be used on signal assignment statements unless the statement appears within a block that has a guard_expression. **guarded** must appear immediately to the right of the signal assignment operator:

```
signal_2 <= guarded signal_1 ;
```

Assignment is made when the value of the GUARD signal changes to TRUE or when one or more of the signals to the right of **guarded** change and the value of GUARD is TRUE. In the example given above the current value of *signal_1* is assigned to *signal_2* when GUARD changes from FALSE to TRUE. If GUARD is TRUE the new value of *signal_1* is assigned to *signal_2* when *signal_1* has an event (i.e. changes value).

The use of the **guarded** option in a concurrent signal assignment statement is equivalent to placing a sequential signal assignment statement in an **if** statement that has GUARD as the 'if' expression.

```
signal_2 <= guarded signal_1 ;
```

is equivalent to:

```
equivalent :    process
                begin
                    if GUARD then
                        signal_2 <= signal_1 ;
                    end if ;
                    wait on  GUARD, signal_1 ;
                end process equivalent ;
```

15.5.5 Guarded Signal Assignment

- **Conditional assignment**

```
guard_example :
        block (clk'EVENT and clk = '1')
        begin
                sig <= guarded in_1 when a = 1 else
                               in_2 when a = 2 else
                               in_3 ;
        end block guard_example ;
```

- **Selected assignment**

```
guard_example :
        block (control = 1)
        begin
            with tech select
                sig <= guarded in_1 when load,
                               in_2 when clear,
                               in_3 when add ;
        end block guard_example ;
```

15.5.5 Guarded Signal Assignment

The examples in Illumination Box 15.5.5 show how the **guarded** option might be used in a conditional signal assignment statement and in a selected signal assignment statement.

The **guarded** option can only be used with concurrent signal assignment statements and must be placed immediately to the right of the signal assignment operator. It is the signal assignment operator that is being "guarded". Think of the **guarded** option as a drawbridge at the entrance to a castle. When the guard condition is TRUE the bridge is down and the value can be assigned to the signal on the other side of the assignment operator. When it is FALSE the bridge is up and the value cannot be passed to the signal.

Note that **guarded** appears only once in the statements. It is not necessary to place **guarded** on each line of the signal assignment statement. It is the operation of assignment which is conditioned by the **guarded** option. All assignments that are made through the guarded operator are subject to the condition.

Once a signal operator is guarded no assignment can be made through that operator unless the value of the GUARD signal is TRUE. In the case of the conditional assignment example on the facing page *sig* is assigned the value of *in_1* only when a rising clock edge has occurred and the value of *a* is 1. In the case of the selected assignment example *sig* is assigned the value of *in_1* only when *control* is equal to 1 and the value of *tech* is *load*.

The conditional signal assignment statement is evaluated whenever there is an event on GUARD, *in_1, in_2, in_3* or *a*. The selected signal assignment is evaluated whenever there is an event on GUARD, *in_1, in_2* or *in_3*.

15.6 Block Declaratives

- Appear between guard expression and begin

- May contain
  - definition of block interface
  - declarations

- Objects declared here are limited in scope to within the block

- Block interface
  - defines specific relationship between block and environment
  - contains port and generic clauses and port and generic maps
  - is an added step towards structure

15.6 Block Declaratives

Like other structures in VHDL blocks have a declarative region between the first line of the block statement and the word **begin**:

```
block_label :   block [(guard_expression)]
                        -- declarative_region
                begin
                        -- concurrent statements
                end block block_label ;
```

This region can be used to define the block interface (header) and to make declarations.

Declarations made in the block declarative region are limited in scope to the block. All declarations that may be made in a concurrent declarative area may be made in the block declarative region. This includes:

1. type declarations
2. subtype declarations
3. constant declarations
4. signal declarations
5. subprogram declarations

The block interface header defines the specific relationship between the block and the rest of the model. It may contain port clauses, generic clauses, port maps, and generic maps. It is a step towards structure. Some aspects of the structure provided by the block header are somewhat artificial. The existence of a header does not limit the visibility of objects declared in the declarative area of the architecture. For example:

```
block_label : block [(guard_expression)]
                generic (de_lay : TIME := 0 ns ) ;
                port (in_1 : in Std_logic ;
                        out_1 : out Std_logic) ;
                generic map (5 ns) ;
                port map (i_put, o_put) ;
        begin
```

15.6.1 Block Attributes

- **Return Boolean values**
- **block_name'BEHAVIOR**

  returns TRUE if block does not contain a component instantiation statement

- **block_name'STRUCTURE**

  returns TRUE if block does not contain a non-passive process statement

15.6.1 Block Attributes

Several kinds of attributes have been introduced and discussed in this book. In addition to signal and array attributes there are also block attributes.

Each of the two block attributes returns a boolean value (TRUE or FALSE) depending upon the result of a simple comparison:

1. block_name'BEHAVIOR
   This attribute returns TRUE if the block to which it is applied does not contain a component instantiation statement. Stated more simply, if the block is coded in a purely behavioural manner this attribute returns TRUE. A component instantiation statement is a structural coding statement. This type of statement and structural modelling is discussed in the next chapter.
2. block_name'STRUCTURE
   This statement returns TRUE if the block does not contain an active process. More simply stated; if there is any behavioural coding in the block this statement returns FALSE. A non-passive process statement could be a concurrent signal assignment statement, a **process** statement which does some type of signal value determination and assignment, or any of the statements that cause something to happen of the kind seen so far in this book.

15.7 Exercises

15.7.1 Create a model of a device which calculates the mean, median, and mode from a line source which supplies integer numbers from 1 to 100. The device accepts numbers from the line when an enable signal is '1'. When the enable signal turns to '0', it stops accepting numbers and calculates the mean, median and mode. When the enable turns to '1' again the device should reset all stored values to zero and begin storing a new set.
The device must keep track of the number of numbers and the numbers themselves during a period of unknown length. When the enable is low, i.e. drops to '0', the device is to begin the calculations. Output the mean, median and mode on three separate lines. NB. Ignore multiple mode situations.

15.7.2 Write descriptions for
a) An octal transparent latch with 3-state outputs (74373) and
b) An octal D-Type flip-flop with 3_state outputs (74374)
The pin names are as follows

D0-D7	Data Inputs
LE	Latch Enable (Active HIGH) Input
CP	Clock (Active HIGH going edge) Input
OE_bar	Output Enable (Active LOW) Input
O0-O7	Outputs

TRUTH TABLES

373

Dn	LE	OE_bar	On
H	H	L	H
L	H	L	L
X	X	H	Z*

374

Dn	CP	OE_bar	On
H	$\uparrow$	L	H
L	$\uparrow$	L	L
X	X	H	Z*

H = HIGH Voltage Level
L = LOW Voltage Level
X = Immaterial
Z = High Impedance
*Note: Contents of flip-flops unaffected by the state of the Output Enable input (OE_bar)

15.7.3 Design a sequence detector having a single input X and a single output Z. In any clock interval in which X = 0 the output is to be Z = 1 provided that in the four previous intervals the input was X = 1001, overlapping to be allowed.

15.7.4 Re-design the 8-bit shift/storage register of Ex. 10.10.3 using guarded block statements.

Chapter 16

Structural Modelling, Generics and Generate

Overview

VHDL structural modelling is primarily a way in which models can be linked together to form a larger model. At gate level it is simply a netlisting exercise for primary gates but at higher levels it is a means whereby gate level, RTL, or behavioural models may be interconnected as components. It is important to remember that, at some point, components in a structural model must have behaviour.
Generics are a method for passing data into a model, the data being restricted to type constant.
A generate statement provides a mechanism for iterative or conditional elaboration of a portion of a description.

In this chapter you will learn

• to construct "pure" structural models

• how to write generic declarations

• how to use the generate statement

16.1 Structural Modelling

- Describes entity in terms of its subcomponents and how they are connected
- Hierarchical arrangement of interconnected components
- Ports and their interconnection are central focus
- Very similar to netlisting in schematic capture systems
- Has to be supported at some level by behavioural description (unless already at gate level)

16.1.1 Structural Steps

- Declare component
- Instantiate component
- Specify which component is associated with each instance

Introduction to VHDL

16.1 Structural Modelling

Pure structural modelling describes an entity in terms of its subcomponents and how these are connected. It is, in effect, a hierarchical arrangement of interconnected components.

Although the following pages discuss "pure" structural modelling it must be kept in mind that, at some level, components must have behaviour or, at least, an architecture which can be simulated, say, at gate or RTL level. They therefore have to be supported by behavioural, or architectural, models. Structural models are simply a way to connect such models together. In some ways a structural model is similar to instantiating the symbols for VHDL behavioural or architectural models on a sheet and connecting the symbols together with nets.

The central focus of a pure structural model is the ports of the subcomponents of the model and their interconnection. This process is somewhat simplified by thinking of signals as "wires" which carry values from one point to another. In this sense the process of structural modelling is much like netlisting.

It should be pointed out that structural and behavioural modelling techniques can be intermixed. If a VHDL behavioural model has already been created its functionality can be included in another behavioural model by instantiating the first model in the second using structural statements. The fact that this "mixed" style of model must exist was hinted at when the **block** attributes 'BEHAVIOR and 'STRUCTURE were discussed.

16.1.1 Structural Steps

There are three steps associated with the construction of a structural model:

1. Component Declaration

   The component declaration identifies a name, generic list, and port list for a "device" used in the structural model. The name of the component, the generics, and ports are "local" names. They are unique to the specific structural model where the declaration is made.

   The process of component declaration is similar to setting up a template for what the component looks like.

   Component declaration can be made in declarative areas of the model or in packages. The advantage of making the declarations in packages is that a consistent declaration is available to all modellers and the need to create a declaration in every model is removed.

2. Component Instantiation

   The component instantiation places the declared component in an activity area of the model and identifies how the signals in the model connect to that particular instance.

   It should be remembered that signals may be thought of as "wires" or

16.2 Component Declaration

- **Describes local interface**
  - **name**
  - **ports and generics**
- **Appears in declarative areas**
- **Format**

```
component identifier_name
    [generic (local_generic_list) ;]
    [port (local_port_list) ;]
end component ;
```

- **Example**

```
component and_gate
    port (a, b : in Std_logic :
              c : out Std_logic) ;
end component ;
```

conduits" for values. The "wire" concept is applicable here. Signals conduct values from component to component within the structural model.

3. Component Identification
   Component identification identifies which design entity is associated with the component.
   This statement appears in the declarative area of the model. It shows the actual name of the entity which is represented by the local name of the entity appearing in the component declaration.

16.2 Component Declaration

Component declarations describe the local interface of components. They provide names for the components and describe the ports and generics associated with the components. Their general format is:

> **component** identifier_name
> [**generic** (local_generic_list) ;]
> [**port** (local_port_list) ;]
> **end component ;**

The identifier name given to the component in the component declaration is a local name. It is not the name of the behavioural model with which the component is associated; however, to clarify the structural model, it should reflect the functionality of the behavioural model. A local component name like *xyz* has no logical association and would make the structural model difficult to decipher for the casual reader.

The identifier name in the component declaration is the name used for the component instantiation statement and the component specification statement.

The **generic** statement and the **port** statement are optional; however, without a **port** statement there is no way in which to transfer values into the component or out of the component. Similarly, without a **generic** statement the system assumes that there are no generics in the model.

The port and generic names appearing in the component declaration are local names. In a sense they are simply place holders which identify the number and placement of ports and generics. These names are not used in the activity area of the structural model and they are not the actual names used for the ports and generics in the behavioural model associated with the component.

The component declaration is made in the declarative area of the model. A declaration for a component must appear before the component specification statement which links the local component name to a specific entity name.

16.2.1 Component Instantiation

- **Places component in relationship to other components**

- **Identifies signals connected to component parts**

- **Associates values to generics**

- **Format**

```
*label : component_declaration_name
            [generic map (association_list) ;]
            [port map (association_list) ;]
```

- **Example**

```
u2 : and_gate
      port map (signal1, signal2, signal3) ;
```

***label is mandatory**

16.2.1 Component Instantiation

Component instantiation refers to the placement of a declared component in an activity area of the model.

The component instantiation statement is a concurrent statement. It cannot be used in sequential areas of the model. The format is:

label : component_declaration_name
    [**generic map** (association_list) ;]
    [**port map** (association_list) ;]

1. A label is mandatory for a component instantiation statement.
2. The component_declaration_name must match the name used in the component declaration statement.
3. The **generic** and **port** map clauses are needed if **generic** and **port** statements appeared in the component declaration statement.
   The purpose of the mapping clauses is to show the relationship between the local name used in the component declaration statement and the actual signal name used in the structural model.
   For example:

   u2 : example_of _association
       port map (a => signal1, b => signal2, c => signal3) ;

   A component may be instantiated several times in the structural model. Different instantiations usually have different signals connected to the component. The port map in this component instantiation statement indicates for this instantiation that the actual signals in the model are *signal1, signal2 and signal3*. `Signal1` is associated with the local name `a` in the component and so on. The port map can also be written:

       port map (signal1, signal2, signal3) ;

   The association in this map is by position rather than by name as in the previous example. Maps identify the actual signals connected to the component ports or generics.

Placing a component instantiation statement in an activity area in the model is similar to placing an instance of the component on a sheet when using schematic capture tools. The **port map** tells the system what signals are connected to the ports of the component declaration.

16.2.2 Component Specification

- **Identifies which design entity is to be used as component**

- **States**
  - **name of entity**
  - **name of architecture**

- **Format**
  for component_instantiation_label :
  component_name **use** binding_indication ;

- **Binding_indication**
  entity library.entity_name
  [(architecture_name)]

- **Example**
  for u2 : and_gate
  use entity WORK.my_component
  (arch_one) ;

16.2.2 Component Specification

Component specification is the process of identifying which design entity is to be used for each instantiation of the component. It is important to realise that different design entities (models) can be used for different instantiations of the same component. It is also possible to use different architectures of the same design entity for different instantiations of the component.

This statement identifies the design entity and architecture that is to be used for a particular component instantiation; it appears in declarative areas of a model and has the format:

for component_instantiation_label : component_name ;
use entity library.entity_name [(architecture_name)] ;

1. There must be a component specification statement for each component instantiation within the model.
2. The component_instantiation_label must match the label used before the component instantiation statement (remember that a label is required on the concurrent component instantiation statement.
3. The entity_name is the name given in the entity declaration of the model. It, along with the component instantiation label, is used to identify a specific instance.
4. The **use entity** clause identifies the particular behavioural (or structural/ behavioural) model that is to be used as the functionality for this instantiation. Three pieces of data are needed to make a positive identification of the behavioural model:
   - Library name

     The library name given here must be a logical name which appears in a **library** statement at the beginning of the model. This name must then be mapped to the directory containing the behavioural model (component) that is to be used for the functionality of the instance. The rules for naming the library and associating the library with a physical structure are the same as those discussed for libraries containing packages.
   - Entity name

     The entity name is the name of the component (and entity, since they are the same) which is to be used for the functionality of the instance. It must be a component in the directory identified by the library name.
   - Architecture name

     The architecture name is optional. If multiple architectures exist for the same entity and the architecture name is not given in the component specification statement the architecture which was compiled last will be used.

16.3 Structural Layout

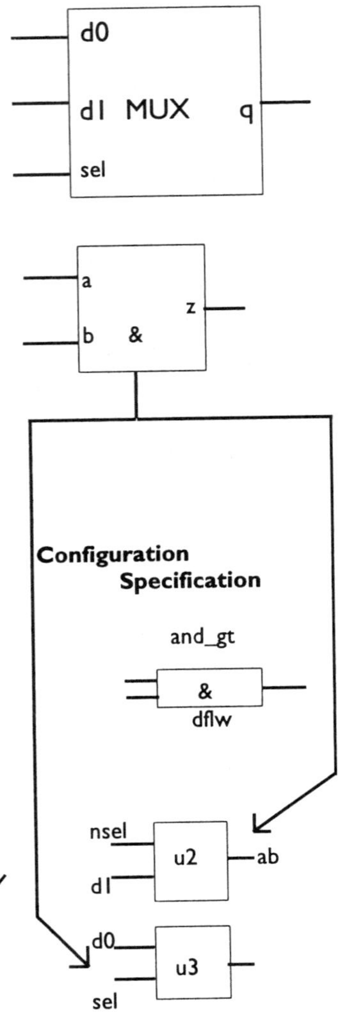


```
entity mux is
  port (d0, d1, sel : in BIT; q out BIT) ;
end mux ;

 architecture structure of mux is
 component and2
    port (a, b : in BIT ; z : out BIT) ;
  end component ;
 component or2
   port (a, b : in BIT ; z : out BIT) ;
 end component
 component inv
   port (i : in BIT ; z : out BIT) ;
 end component ;          -- component to be
                          -- used in mux
 signal aa, ab, nsel : BIT ; -- internal signals

 for u1 : inv use entity invrt(behav) ;
 for u2, u3 : and2 use entity and_gt(dflw) ;
 for u4 : or2 use entity or_gt(arch1) ;
 -- declares components and architectures to use
-- from package
begin
  u1 : inv port map ( sel, nsel) ;
  u2 : and2 port map (nsel, d1, ab) ;
  u3 : and2 port map ( d0, sel, aa) ;
  u4 : or2 port map ( aa, ab, q) ;
 end structure ;
```

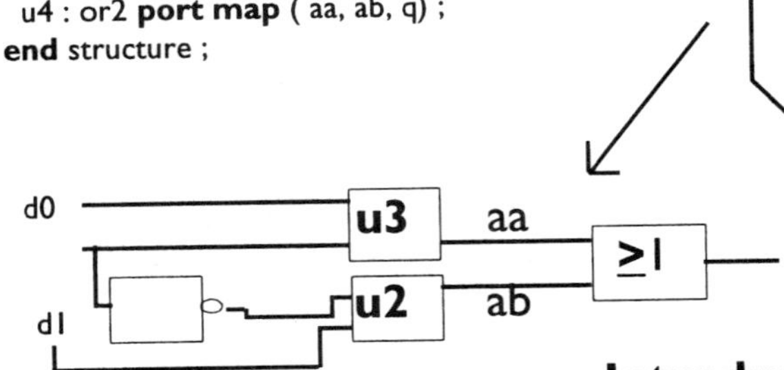

Component Instantiation

Introduction to VHDL

16.3 Structural Layout

The example shown in Illumination Box 16.3 is illustrative in explaining the steps in structural modelling.

The port names *d0, d1, sel* and *q* are the signals entering the design and are used in this model. The signals *aa, ab,* and *nsel* are intermediate signals used within the design to connect components to each other. Three components are used in this model; an and_gate called AND2, an or_gate called OR2 and an inverter called INV. These components appear in component declaration statements.

The and_gate is instantiated twice in the activity area of the model and the inverter and or_gate are both instantiated once. The statement labels chosen for the instantiations are *u1, u2, u3* and *u4*. However, any label chosen by the user would be acceptable. The **port map** in each of the instantiations shows what signals in the model are connected to the ports in the component declarations. Notice that some internal, and external, signal names appear in more than one port map. A study of the schematic shows that an internal signal which receives an output from one instance may be connected to the input of another instance. Thus the name of the signal must appear in two, or more, port maps.

The component specification statements identify the actual behavioural models that are used for each of the instances which appear in the activity area of this model. Notice that the two labels *u2* and *u3* appear in one specification statement. This indicates that both instances derive their functionality from the same behavioural model. Also notice that the name of the component in the **use entity** clause is not preceded by a library name. This is allowable when the structural model and the behavioural models are in the same directory (working directory or WORK).

16.3.1 Structural Example

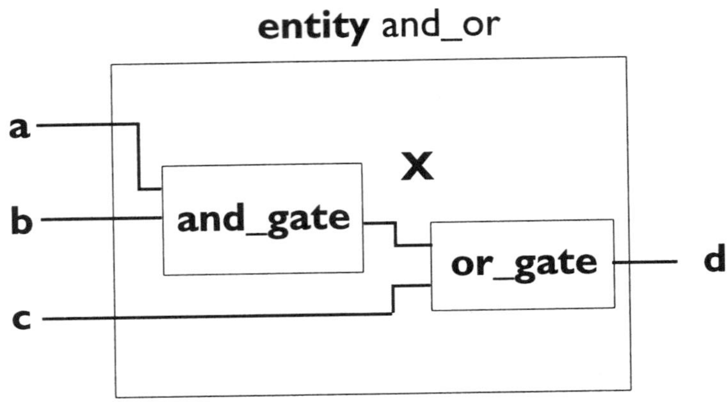

- **Specification**
  Create model of entity with three input ports (a, b, c) and one output port (d). Signals a and b are anded and the result X is ored with signal c to produce output signal d.

- **Analysis**
  Two components will be declared, specified and instantiated. One for the and_gate and the other for the or_gate. In addition, an internal signal X, which represents the output of the and_gate and input of the or_gate will be declared.

16.3.1 Structural Example

The example shown in Illumination Box 16.3.1 gives a specification of a device and the analysis of how the device could be modelled structurally.

The signal names *a, b, c, x* and *d* are the actual names to be used in this model. The signals *a, b* and *c* are input signals, *d* is an output signal. These names appear in the **port** statement in the entity declaration. The signal *x* is an internal signal and is declared in the declarative area of the architecture.

Since behavioural models of both the *and_gate* and the *or_gate* have already been created the model can be constructed structurally by declaring two components (one for the *and_gate* and one for the *or_gate*), identifying the behavioural models which are associated with the structural instantiations, and instantiating the components.

The completed structural model is shown overleaf.

16.3.2 Structural Model

```
library in_house ;
library IEEE ;
use IEEE.Std_logic_1164.all ;

entity and_or is
        port (a, b, c : in Std_ulogic ;
                    d : out Std_ulogic) ;
end and_or ;

architecture structure of and_or is
        component and1
                port (x, y : in Std_ulogic ;
                         z : out Std_ulogic) ;
        end component ;
        component or1
                port (x, y : in Std_ulogic ;
                         z : out Std_ulogic) ;
        for u1: and1
                use entity in_house.and_gate (beh)
                port map (a, b, c ) ;
        for u2 : or1
                use entity WORK.or_gate (beh) ;
        signal x : Std_ulogic ;
begin
        u1 : and1 port map (a, b, x) ;
        u2 : or1 port map (x, c, d) ;
end structure ;
```

16.3.2 Structural Model

The model in Illumination Box 16.3.2 shows how the specification of the device depicted in Illumination Box 16.3.1 could be constructed structurally.

1. The entity declaration is constructed in the same manner whether it is used with a structural model or a behavioural model.
2. In the declarative area of the architecture the components are declared and the behavioural models to be used in each instantiation are specified.
3. The component declarations are made first. Notice that the port names are the same in each of the declarations. This is allowable because the names are place holders. However, the number of names must match the number, mode and type in the behavioural models which are linked to the component instantiations.
4. The component specifications are made after the component declarations. Notice that the name used in the specifications consists of both the instantiation statement label and the component declaration name. Both are needed to identify the instantiation. It should be remembered that a component can be instantiated several times and that a component specification is needed for each instantiation.

   Note the use of a **library** statement for "in_house" at the beginning of the model. Observe that the name (in_house) appears after the **use entity** for u1 and and1. This indicates that the behavioural model used for this instance is in a library other than **work**. Also notice the use of a **port map** in the component specification statement. This is necessary when the port names in the component declaration are not the same as the port names in the model specified in the component specification statement. The **port map** shows which ports in the component declaration are associated with which ports in the behavioural model (in this case positional association is used).
5. The internal signal *x* is also declared in the declarative area of the architecture. This signal "carries" the result from the *and_gate* to the *or_gate*.
6. The two components are instantiated in the activity area of the architecture.

   Component instantiation statements are concurrent statements. Unlike some of the other concurrent statements they must have a user supplied label. This label is needed to identify the instantiation in the component specification statement.

   The port map in each of the instantiations shows which actual signals are connected to the component. In both cases the association of actual

16.4 Generics

- **Provide a method for passing additional data into model**
- **Are type constant of mode in**
- **Declared in interface lists in**
  - entity declaration
  - component declaration
  - blocks
- **Data is assigned to generics from**
  - symbol properties
  - generic mapping statements

to local is done by positional notation. For *u1*, input signal *a* connects to *x*, *b* to *y*, and *x* to *z*. In this last case the *x* is the *x* declared in the model and not the "place" holding *x* in the component declaration. For *u2*, signal *x* is connected to *x*, *c* to *y*, and *d* to *z*. It should be recognised that *d* is the output signal specified in the entity port statement.

16.4 Generics

In Chapter 14 packages were presented as a method of passing information to models. Another method for passing additional data (restricted to the object type **constant**) into a model is provided by the VHDL construct called **generic**.

Generics are declared in generic interface lists which can appear in:

1. Entity declarations
2. Component declarations
3. Blocks

Generics are constants of mode **in** which derive their values from one of two sources.

1. Symbol properties
   Models which are linked to symbols can use generics to access values assigned to properties assigned to the symbols.
2. Generic mapping statements
   A statement called a generic mapping statement can be placed in models. This statement can be used to assign values to generic constants which appear in the generic interface lists

A default value can appear in the generic interface list. The generic assumes the default value when neither a symbol property nor a generic mapping statement exists for the generic.

16.4.1 Generic Statement

- **Declares generics**
- **Precedes port declaration**
- **Format**

```
generic (class generic_name :
                    mode type [default_value] ) ;
```

- **Example**

```
generic ( constant de_lay : in TIME := 5 ns) ;
or
generic (de_lay : TIME := 5 ns) ;
```

- **Use**

```
entity generic_use is
        generic (de_lay : TIME := 5 ns) ;
        port ( in_a : in BIT ; out_b out BIT) ;
end generic_use ;

architecture example of generic_use is
begin
        out_b <= in_a after de_lay ;
end example ;
```

16.4.1 Generic Statement

Generic constants or, more simply, generics are declared in **generic** statements. These statements can appear in entity declarations, component declarations, or block headers.

When **generic** *statements appear in a model they must precede the* **port** *statement.*

The general format is:

generic (**constant** name : **in** type [:=default_value]) ;

Because all generics must be constants of mode **in** this format is usually expressed as:

generic (name : type [:= default_value]) ;

Examples of generic statements might be:

generic (**constant** de_lay : **in** TIME := 5 ns) ;
generic (ti_me : TIME) ;

Although the default value in the **generic** statement is optional, failure to use it may result in an error. Generics must be assigned a value; this value may be assigned in the generic statement, or in a generic mapping statement, or as the value of a symbol property. If a specific value is not provided, an error results. Unlike other VHDL objects (constants, variables, signals) generics do not assume the left-most value of their type when no value is provided. Because there is no actual "default" value, the value in the generic statement might be thought of as an "initial" value rather than a "default" value. If this initial value is not provided and there is no assignment of a value, an error results.

16.4.2 Mapping Generic Values

- Assigning generic values through mapping
- Values may be assigned to identifiers in generic statement through generic map
- Format:

```
generic map (value_1, value_2, . . ) ;
```

- Each value in map must be associated with an identifier in generic list
- generic maps may appear in component specifications, component instantiations and block headers
- Example:

```
u2 : and_gate
     generic map (3 ns)
     port map (signal1, signal2, . .) ;
```

16.4.2 Mapping Generic Values

Values may also be assigned to generic identifiers by generic mapping statements. These statements can appear as block headers, component specification statements, or component instantiation statements.

The generic mapping statement can be used to assign values to generic identifiers appearing in generic statements or to show association between formal generics and local generics in a component declaration statement. The general format is:

generic map (value_1, value_2, . . .) ;

Values in the map may be associated with identifiers by positional, or named, association. Each value in the map may be associated with an identifier in the generic statement. There cannot be unassociated values in the generic map statement.

An example of the use of a generic map statement in a component instantiation statement is:

u2 : and_gate
  generic map (3 ns)
  port map (signal1, signal2, signal3) ;

In this example a component called and_gate has been instantiated in the model. The behavioural model associated with this component must have a generic statement which has one identifier in its entity (and a **port** statement with three port identifiers). The generic map in this component instantiation statement associates the value 3 ns with the generic identifier in the behavioural model. When the behavioural model is executed the value 3 ns is used wherever the generic identifier appears in the model.

It is also possible to use the reserved word **open** to indicate that the default values of the generic values specified in the declaration are to be used, viz:

generic map (open, value_new, . . .)

Note that value_2 has been assigned the new value *value_new* as being a value not specified by its default value in the list.

Finally **open** can be used with named associations in the same way as with port maps (See Section 16.2.1) but it should be noted that if **open** is used with port maps this value indicates an open signal (wire) rather than a default value.

16.4.3 Generic Use Example

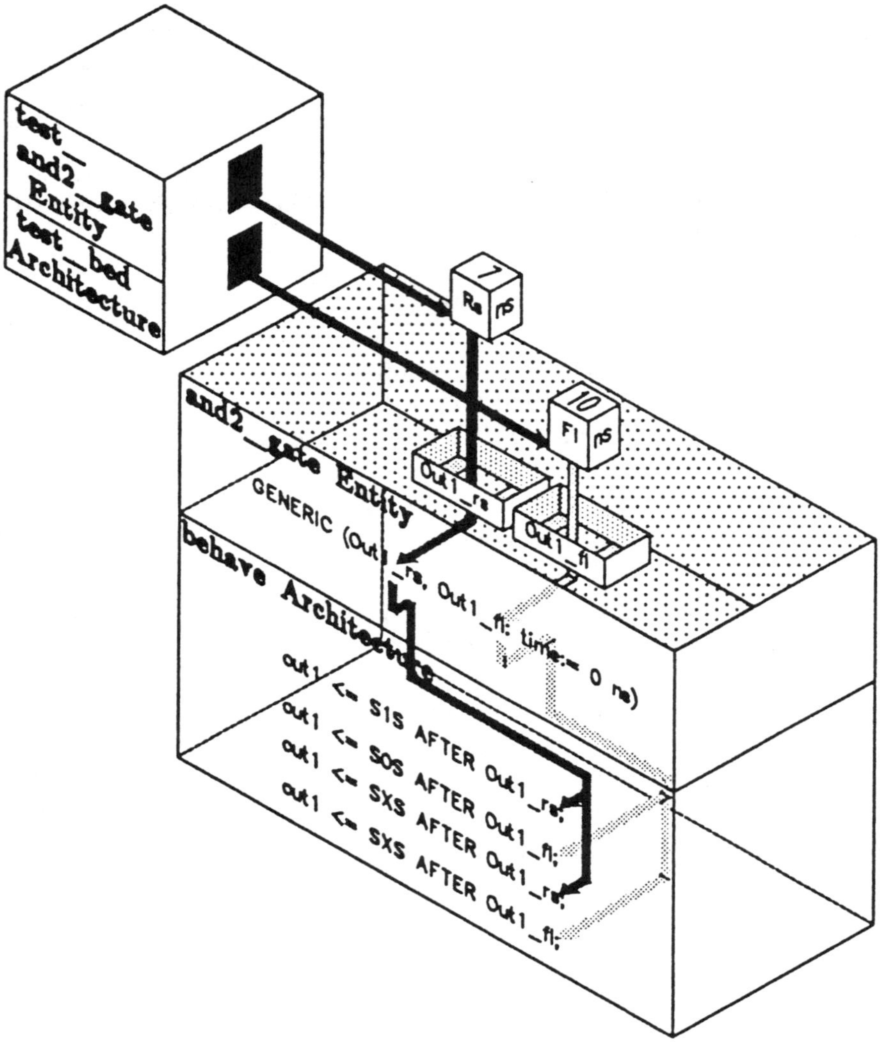

16.4.3 Generic Use Example

The model in the lower (larger) box in Illumination Box 16.4.3 represents a behavioural model. A **generic** statement appears in the entity declaration of this model. Notice that this statement declares two generic constants of type TIME and provides a "default" value of 0 ns for each of the constants. If this model were to be compiled as a stand-alone VHDL model the delay assigned by the **after** clause in the signal assignment statements would be 0 ns.

The model shown in the upper (smaller) box represents a structural model. This model has a component specification statement which identifies the lower model as the functionality for one or more of the instantiations in its architecture. In addition, a **generic map** statement appears in the instantiation statement which maps the value 7 ns to *out1_rs* and 10 ns to *out1_fl*.

When the structural model is simulated the value 7 ns is used in the behavioural model whenever *out1_rs* is encountered; similarly, 10 ns is used whenever *out1_fl* is encountered.

16.5 Configuration Declarations

- **Provides a mechanism for specifying deferred bindings of component instances**

- **Binding of component instances is performed by configuration specifications**

- **Example**

```
architecture structure of RISC is
component register_bank    (...) end component ;
component alu              (...) end component ;
component main_control     (...) end component ;
component memory           (...) end component ;
begin
P1: register_bank       port map (...) ;
P2: alu                 port map (...) ;
P3: main_control        port map (...) ;
P4: memory              port map (...) ;
end structure ;
library PU_RISC, WORK ;
configuration preferred_version of  RISC is
use WORK.all :
for structure  -- configuration specification
   for P1 : register_bank
         use configuration PU_RISC.Instr_set_1 ;
   end for ;
```

16.5 Configuration Declarations

It has been shown how a design entity may be described in terms of a hierarchy of blocks (Chapter 15) and also how it may be described in terms of unconnected components. We have seen how each component of a design entity can be bound to a lower level design entity in order to define the structure, or behaviour, of that component. Thus, a design entity can be successively decomposed into components and these components may be bound to other design entities which themselves may be decomposed in a similar way. The consequence is a hierarchy of design entities which represents a complete design. Such a collection of design entities is described in VHDL as a design hierarchy. The bindings necessary to identify a design hierarchy can be specified in a configuration of the top level entity in the hierarchy. Such a configuration is defined by a configuration declaration.

The binding of component instances to design entities is performed by configuration specifications, and such specifications appear in the declarative part of the block in which the corresponding component instances are created. Sometimes, however, it may be preferred that specification of the component binding in a given block be deferred until later. A mechanism for specifying such deferred bindings is provided by the configuration declaration.

For example, in the exercise 16.8.5 the architecture of the RISC processor might be summarised as follows:

```
architecture structure of RISC is
component register_bank (...) end component ;
component alu           (...) end component ;
component main_control (...) end component ;
component memory        (...) end component ;
begin
P1: register_bank port map (...) ;
P2: alu           port map (...) ;
P3: main_control port map (...) ;
P4: memory        port map (...) ;
end structure ;
library PU_RISC, WORK ;
configuration preferred_version of RISC is
use WORK.all :
for structure -- configuration specification
    for P1 : register_bank
          use configuration PU_RISC.Instr_set_1 ;
    end for ;
```

•

16.6 Generate Statement

- **Provides a mechanism for the elaboration of a portion of a description**

- **Available modes**
  - **Iterative**
  - **Conditional**

- **Format**

generate statement ::=
    *generate*_label :
        generation_scheme **generate**
            [{ block_declarative_item }
        begin]
            { concurrent_statement }
        end generate [*generate*_label] ;

generation_scheme ::=
    for *generate*_parameter_specification
    | **if** condition
    label ::= identifier

16.6 Generate Statement

There are many instances in digital circuit design where elements of the design may be repeated identically. One way of realising such repetitions has already been discussed in the sections on structural modelling and configurations declarations.

The **generate** statement provides another mechanism for iterative or conditional elaboration of a portion of a description.

The general format is :

generate statement ::=
*generate*_label :
generation_scheme **generate**
[{ block_declarative_item }
begin
{ concurrent_statement }
end generate [*generate*_label];
generation_scheme ::=
for *generate*_parameter_specification
| **if** condition
label ::= identifier

If a label appears at the end of a generate statement it must repeat the generate label.

For a **generate** statement with a **for** generation scheme the generate parameter specification is the declaration of the *generate parameter* with the given identifier. The generate parameter is a constant object whose type is the base type of the discrete range of the generate parameter specification.

The discrete range in a generation scheme of the first form must be a static discrete range; similarly, the condition in a generation scheme of the second form must be a static expression.

The **for** generation scheme is similar to the **for loop** statements described in Section 10.3 and the **if** generation scheme is similar to the **if** statement described in Section 8.2.1.

16.7 Generate Example

```
------------------
-- n bit register
------------------

        entity shift_register is
            generic (reset_level : Std_ulogic := '1' ;
                        reg_stages : INTEGER := 2#1011#) ;
        port ( input, clock, reset : in Std_ulogic ;
                    output : out Std_ulogic ;
        end shift_register ;

        architecture shift_register_arch of shift_register is
        component register_cell
            generic (reset_level : Std_ulogic ) ;
            port (input, clock, reset : in Std_ulogic ) ;
        end component ;
        constant clock_edge : Std_ulogic ) ;
        signal z : Std_ulogic_vector (reg_stages downto 0) ;
        begin
            gen : for i in 0 to (reg_stages - 1) generate
                    regx : register_cell
                            generic map (reset_level) ;
                            port map (z(i), clock, reset, Z(i + 1) ;
            end generate gen ;
            z(0) <= input ;
            output <= z(reg_stages) ;
        end shift_register_arch ;
```

16.7 Generate Example

The example on the facing page is the description of an *n* bit shift register which uses a previously designed shift register cell for its elaboration.

The reset_level has been declared as a generic in order that both active high and active low versions are possible

Reg_stages has been defined in binary digit form and, of course, may be varied for any length register.

Within the architecture a clock_edge has been declared as a CONSTANT in order that it may be redefined for an active low clock.

The connection between stages is achieved by the use of the internal signal z and the generate statement labelled *gen* creates the n stages of the register. The final concurrent statements for *z(0)* and *output* connect these to their respective ports.

16.8 Exercises

16.8.1 Create a complete model for the Scintillation Detector of Chapter 3 using:

- a) The implicit package WORK.
- b) The constructs of **function** and **procedure.**
- c) Packages.

16.8.2 Design a complete system for the filter of Exercise 13.1. NB. Some elements related to this problem are to be found among the exercises in Chapter 9.

16.8.3 Create a model for an n-bit serial input shift register which can be modified for active high , or active low, clock and reset inputs.

16.8.4 a) Write a description for an n-bit cyclic redundancy checker which is modifiable to most of the standard, general purpose, CRC polynomials.

b) Create a model for an IBM 16-bit CRC (Cyclic Redundancy Check) generator in which the constant "divisor" is given in binary as:

1000 1000 0001 0000 1

or in "quadratic" form as:

$G(X) = X^{16} + X^{12} + X^{5} + 1$

The process is begun by adding 16 zeros (1 fewer bits than the number of bits in the "divisor") after the LSB of the message to be sent. The "divisor" is exclusive **or**ed with the 16 most significant bits of the message, leading zeros having been ignored. Enough bits from the message to be sent are appended to the result of this operation to form another 16 bits of data again headed by a 1. This exclusive **or**ing is repeated along with the rest of the process until all of the bits in the message are exhausted. The final exclusive **or** result (the 'remainder') is the set of CRC characters. Enough leading zeros are appended to the CRC characters to form a total of 16 bits.

16.8.5 Design a RISC processor with the following attributes:
The microprocessor should have a 16 bit structure; it should access and compute only 16 bit words. Any instruction should be encoded in a single 16 bit word and internal registers should be 16 bit. Addresses should also be 16 bit, implying access to 2^{16} x 16 bits i.e. 128 kB external memory. Create a 4 bit operation code which matches that of the Motorola MC680X0 microprocessor condition codes.

Appendix A Glossary of Terms

VHDL Reserved Words and Standard Identifiers - with definitions.

NB. For readability the reserved words appear in lower case boldface. They are reserved for significance in the language. The meanings given below may not be ascribed to the IEEE.

abs	Miscellaneous operator which returns absolute value of the operand which is to its right. Predefined for any numeric type.
access	A value of an access type may designate an object created by an allocator
ACTIVE	**signal**'ACTIVE returns value TRUE or FALSE if activity occurred in present delta time period.
after	Inertial delay is modelled with the **after** clause.
alias	Alternative name for an object or part of an object.
all	In <**use** package.**all**> the .**all** indicates that all of the definitions and declarations in this package are to be linked into this model. Used with **library**.
and	Logical operator.
architecture	Architecture Body: establishes relationship between inputs and outputs of design.
array	Collections of values all belonging to the same object. Individual values stored within separate elements within object. Individual elements identified by an index value.
assert	Checks condition and reports message with severity level if condition is not true.
ASSERTION	Concurrent statement; used to make up model's architecture. Sequential operator.
attribute	An attribute defines some characteristics of a named entity.
BASE	Attribute type, predefined; **type**'BASE can be used only in conjunction with another attribute type; **subtype**'BASE used to find information about the subtype's base type.
begin	Sequential operator.
BEHAVIOR	<Block_name'BEHAVIOR> returns TRUE if block does not contain a component instantiation statement.
BIT	Data type, predefined; scalar - enumeration type, does not require a type definition statement.
BIT_VECTOR	Data type; defines an array of bits.
block	Concurrent statement; used to make up model's architecture. Allows grouping of concurrent statements into logical units. Blocks can be nested to further support partitioning of model.
body	Used with subprograms and packages. Specifies the execution of a subprogram. A package body defines the bodies of subprograms declared in the interface to the package.
BOOLEAN	Data type, predefined; logical, scalar - enumeration type, does not require a type definition statement.
buffer	Port Mode: Identifies direction of data flow through port. **buffer** - in or out, limited to one 'in' source.
bus	Interface type. Guarded signal type.
case	Controls execution of one or more sequential statements. Sequential operator; conditional control statement. Used with **is**.

CHARACTER	Data type, predefined; scalar - enumeration type, does not require a type definition statement.
CLASS	Classifies object_name e.g. **signal, variable**.
COMPILE	Compiles source code.
component	Design unit; container (object) for model of design.
configuration	Describes how component instances in a given block are bound to design entities, in order to describe how design entities are put together to form a complete design.
constant	Object assigned an initial value which cannot be changed.
DELAYED	**signal**'DELAYED returns a signal delayed (t) time units.
disconnect	Used to define the time delay to be used in the implicit disconnection of drivers of a guarded signal within a guarded signal assignment.
downto	See RANGE and REVERSE_RANGE.
else	Part of **if - else** construct; provides selection control between two groups of statements based on a single condition.
elsif	Provides multiple alternatives based upon multiple conditions.
end	Step command; executes all iterations in current time step.
ENDFILE	Function for locating the end of files. See TEXTIO.
end if	Termination of **if** statement.
entity	Basic modelling unit is called a Design Entity; it may represent an entire system, a board, a chip or a gate. Entity Declaration: Names entity and defines interface between entity and its environment
ERROR	Severity level - task completed, result wrong.
EVENT	Step command; executes all iterations and advances time to next time step in which statements activate. **signal**'EVENT; returns value TRUE or FALSE if event occurred in present delta time period. clk'EVENT **and** clk = 1 identifies rising clock edge.
exit	Sequential operator. Used to break out of loops. Only occurs within loops.
FAILURE	Severity level - task not completed.
FALSE	Returned according to correct relational operator comparison.
file	A file declaration creates a file object. File types are used to define objects representing files in the host system environment.
for	**for loop**; statements executed once for each value in loop parameter's range. Used with **wait, in.**
function	Provides a method of performing complex algorithms. Must contain a **return** clause. Invoked by an expression. Produces a single return value. Must contain a **return** clause.
generate	A generate statement provides a mechanism for iterative or conditional elaboration of a portion of a description.
generic	Describes local interface. Provides method for passing additional data into model. Is type **constant** of mode **in**.
group	Named collection of entities.
GUARD	Appears in parentheses after **block** statement.
guarded	Guards the assignment operator and prevents the waveform assignment when the value of the GUARD signal is FALSE.
HIGH	Attribute type, predefined; type'HIGH returns largest value in type's value set.
if	Sequential operator; provides conditional control of sequential statements. Condition in statement must evaluate to a Boolean value. Terminates with **end if.**
impure	Function that may return a different value each time it is called.
in	Port Mode: Identifies direction of data flow through the port. **in** - flow is into entity. PINTYPE property.

inertial	Default delay mode for signal assignment statements.
inout	Port Mode: identifies direction of data flow through port. **inout** - flow may be either in or out. Internal signal.
INTEGER	The only predefined integer type. Range is implementation_dependent but is guaranteed to include the range -2147483647 to 2147483647. It is defined with an ascending range.
INTO	Step command; executes a queued statement.
IO	PINTYPE property.
is	Used with **in** i.e. < FILE example **is in** "location";> to identify files. Files are declared in declarative regions. Used with **type** i.e. <**type** example **is**>. Used with **entity** i.e. <**entity** example **is**>.
ITERATION	Step command; executes all initially 'queued' statements.
IXO	PINTYPE property.
label	Identifier. Used with **generate.**
LAST_ACTIVE	**signal**'LAST_ACTIVE returns amount of time since last activity.
LAST_EVENT	**signal**'LAST_EVENT returns amount of time since last event.
LAST_VALUE	**signal**'LAST_VALUE returns value equal to previous value.
LEFT	Attribute type, predefined; type 'LEFT returns left-most value in type's value set e.g. 'LEFT returns zero of BIT_VECTOR (0 TO 7).
LEFTOF	Attribute type, predefined; type 'LEFTOF(x) returns value in type's value set which is to the left of the stated value(x).
LENGTH	E.g. 'LENGTH returns 8 of BIT_VECTOR(0 TO 7).
-LIB	Switch for compiling libraries other than STANDARD. Used in packages.
library	Provides storage for compiled design units. Provides a logical name used to point to the location of a desired package. See **use**.
linkage	Port Mode: Identifies direction of data through port. **linkage** - documentation only, does not pass signal information, cannot be read or written.
LINE	A LINE is a pointer to a STRING value. **type** LINE **is access** STRING. See TEXTIO.
literal	A value that is directly specified in the in the description of a design.
loop	Sequences of statements that are executed repeatedly. Iterative control statement. Controls sequential code.
LOW	Attribute type, predefined; type'LOW returns smallest value in type's value set, e.g. 'LOW returns 0 of BIT_VECTOR(0 TO 7).
map	Used with **port** or **generic** to associate actuals with the formals of the design entity interface implied by the immediately enclosing binding indication.
-MAP	Switch for compiling libraries other than STANDARD. Used in packages.
mod	Multiplying operator - modulus.
MODE	Identifies the direction of data flow with regard to the producer.
MODEL	Property associated with body of symbol.
nand	Logical operator.
NATURAL	Data type, predefined. Subtype of the type INTEGER. Cannot be negative, + **and** * defined, no additive inverse. Range 0 to INTEGER'HIGH.
new	Allocator, the evaluation of which creates an object and yields an access value that designates the object. Also used in association lists as an actual.
next	Causes execution to jump to the beginning of the loop and start the next iteration. Sequential operator. Used to cause completion of a loop iteration.
nor	Logical operator.
not	Logical operator. Inverter switch type BIT.

NOTE	Severity level - general information
null	Sequential statement indicating no action occurs. Sequential operator.
OBJECT_CLASS	Declaring objects: The general format for declaring objects is; OBJECT_CLASS identifier : **type** [:= init_val] ;. Objects; containers for values. (See **constant, variable, signal.**)
of	Used with **is** e.g. <**architecture** example **of** iterations **is**>.
on	Used with **wait**.
open	Used with entity aspect, which identifies a particular design entity to be associated with component instances. See configuration
or	Logical operator.
others	Indicates all unspecified elements; used with **with** a_example **select**.
out	Port Mode: Identifies direction of data flow through port. **out-** flow is out of entity. PINTYPE property.
OVER	Step command; executes a complete 'queued' statement.
package	Shareable collection of declarations made outside models.
PINTYPE	Property associated with pins.
port	Port statement: Identifies ports used by the entity to communicate with its environ ment.
POS	Attribute type, predefined; type'POS(x) returns position number of stated value (x) in type's value set.
POSITIVE	Data type, predefined. Subtype of the type INTEGER. Cannot be negative, + **and** * defined, no additive inverse. Range 1 to INTEGER'HIGH.
postponed	Accesses the 'final' values of signals at the current simulation time.
PRED	Attribute type, predefined; type'PRED(x) returns value whose position number is one less than the position number of the stated value (x).
procedure	Defines how procedure behaves. Consists of sequential statements.
process	Concurrent statement; delineates set of sequentially executed statements.
pure	Function that returns the same value each time it is called.
QUIET	**signal**'QUIET (t) returns a signal value TRUE or FALSE based on activity in (t) time units.
range	specifies a subset of values of a scalar type.
RANGE	Data Type: Scalar; predefined, does not require a type definition statement. Identifies subset of values e.g. 'RANGE returns 0 to 7 of BIT_VECTOR(0 TO 7). May be used with type declarations or object declarations. See VECTOR, **to**.
READ	Procedure for reading from lines. See TEXTIO
READLINE	Procedure for reading from files. See TEXTIO.
REAL	Data Type: Scalar; predefined, does not require a type definition statement. Guaranteed to include the range -1E38 to 1E38. Defined with an ascending range.
record	Composite type, objects of which consist of named elements. The value of a record object is a composite value consisting of the values of its elements.
register	One kind of guarded signal. Retains its last driven value when all of its drivers are turned off.
reject	The time expression following **reject** specifies the pulse rejection limit.
rem	Multiplying operator - remainder.
report	Used with WARNING.
return	Sequential operator; step command; returns evaluation from a subprogram to the main program. Returns value created by a function to the object assigned to receive the object's output.
REVERSE_RANGE	Data type; e.g. returns 7 **downto** 0 **of** BIT_VECTOR(0 TO 7).

RIGHT	Attribute type, predefined; type'RIGHT returns right-most value in type's value set, e.g. 'RIGHT returns 7 **of** BIT_VECTOR(0 TO 7).
rol	Shift left rotate.
ror	Shift right rotate.
RIGHTOF	Predefined type attribute; type'RIGHTOF(x) returns value in type's value set which is to the right of the stated value (x).
select	Concurrent signal assignment; used with **with**. Selection is based on value of included expression.
severity	Used to define WARNING level.
SEVERITY_LEVEL	Data Type: Scalar; predefined, does not require a type definition statement. Enumeration type. See NOTE, WARNING, ERROR, FAILURE.
signal	Object assigned a value with associated time factor which may be changed.
shared	Declares a shared variable.
STABLE	**signal**'STABLE (t) returns a value TRUE or FALSE based on event in (t) time units.
STANDARD	IEEE Std 1076-1987 predefined package. Automatically included in every model.
STD	Library logical name denoting the design library in which packages STANDARD and TEXTIO reside, the only standard packages defined by the language.
STEP	Evaluates model according to specified option.
STRING	Data type, predefined.
STRUCTURE	<Block_name'STRUCTURE> returns TRUE if block does not contain a non-passive process statement.
subtype	Subset of specified type. See **type**.
SUCC	Attribute type, predefined; type'SUCC(x) returns value whose position number is one greater than the position number of the stated value (x).
TEXT	A file of variable-length ASCII records. **type** TEXT **is file of** STRING.
TEXTIO	IEEE Std 1076-1987 predefined package providing textual input and output capabilities. See STD.
then	Conditional signal assignment. Used with **if.**
TIME	Data Type: Scalar; predefined, does not require a type definition statement.
to	See **range, for, in**.
TRANSACTION	**signal**'TRANSACTION returns an event whenever there is an activity on the signal.
transport	Passes signals regardless of duration; used when signals are delayed but 'filtering' effect produced by **after** clause is unwanted. Used with **after**.
TRUE	Returned according to correct relational operator comparison.
type	Can be - data; arithmetic, misc, relational, logical operator, scalar. Identifies the value set which the object can contain. Determines the values an object may assume and the operations which may be performed on it.
unaffected	A waveform in a concurrent signal assignment statement that does not affect the driver of the target.
units	Primary; entity declarations, package declarations, configuration declarations. Secondary; package bodies, architecture bodies. Scalar-physical type e.g. cc ps.
until	Used with **wait**.
use	Makes identifiers in packages visible. See **library**.
VAL	Attribute type, predefined; type'VAL(x) returns value in type's value set which is located at the stated position number (x).
variable	Object assigned a single value which may be changed.
VECTOR	BIT_VECTOR is an unconstrained array. Used with NATURAL and **range**.
wait	Sequential operator. Used with **on, until** and **for**.
WARNING	Severity level - undesirable condition.

when	Required with selected signal assignment statements and CASE statements. Used with **else** as a conditional signal assignment statement and with WITH a_example SELECT.
while	**while loop**; Execution of statements within loop controlled by Boolean equation.
with	Selected signal assignment; used with **select**.
WORK	The working library corresponds to the working directory and has the logical name
WRITE	Procedure for writing to lines. See TEXTIO.
WRITELINE	Procedure for writing to files. See TEXTIO.
xnor	Logical operator
xor	Logical operator.
xnor	Logical operator.
X'DRIVING_VALUE	From within a process, returns the value currently being driven by the process onto the specified signal
X'ASCENDING[N]	Returns TRUE if the if the specified range is ascending
Y'IMAGE(X)	Returns a string containing the string representation of the value X of type T. Useful for displaying enumerated type values
T'VALUE(X)	Returns the value type T represented by the string X. Useful for converting a string to an enumerated type value

*	Multiplying operator - multiplication
**	Miscellaneous operator - exponentiation.
/	Multiplying operator - division.
+	Adding operator - addition.
-	Adding operator - subtraction.
+	Sign operator - Identity.
-	Sign operator - Negation
& single	Adding operator. Used for concatenation, e.g. linking two or more items into a unit. Items must be of the same type.
=	Relational operator. Equal
<	Relational operator. Less than.
>	Relational operator. Greater than
/=	Relational operator. Not equal.
<=	Relational operator. Less than or equal. Only assignment operator which can be used to assign waveforms to signals.
>=	Relational operator. Greater than or equal.

LCS-24: LRM 1993, 7.2.3 : provides 6 new shift operators for bit_vectors. These operators are 3 letter keywords.

sla	Shift right logical
sll	Shift left logical
sla	Shift left arithmetic
sra	Shift right arithmetic
rol	Rotate left logical
ror	Rotate right logical

NB: Commenting Code: A double hyphen (--) indicates that all material from that point to the end of the line is to be treated as a comment.

Appendix B Language Construct Tree

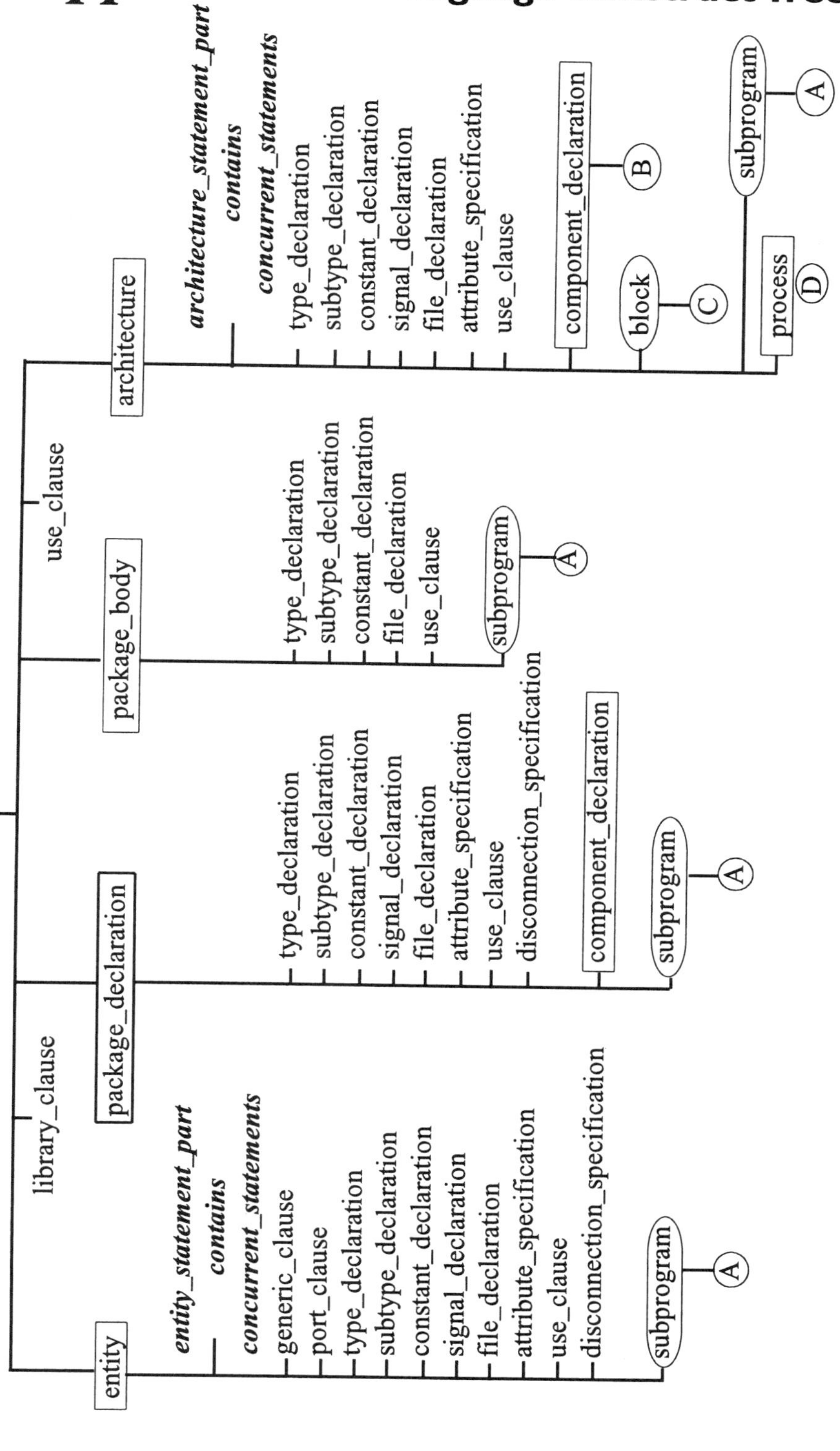

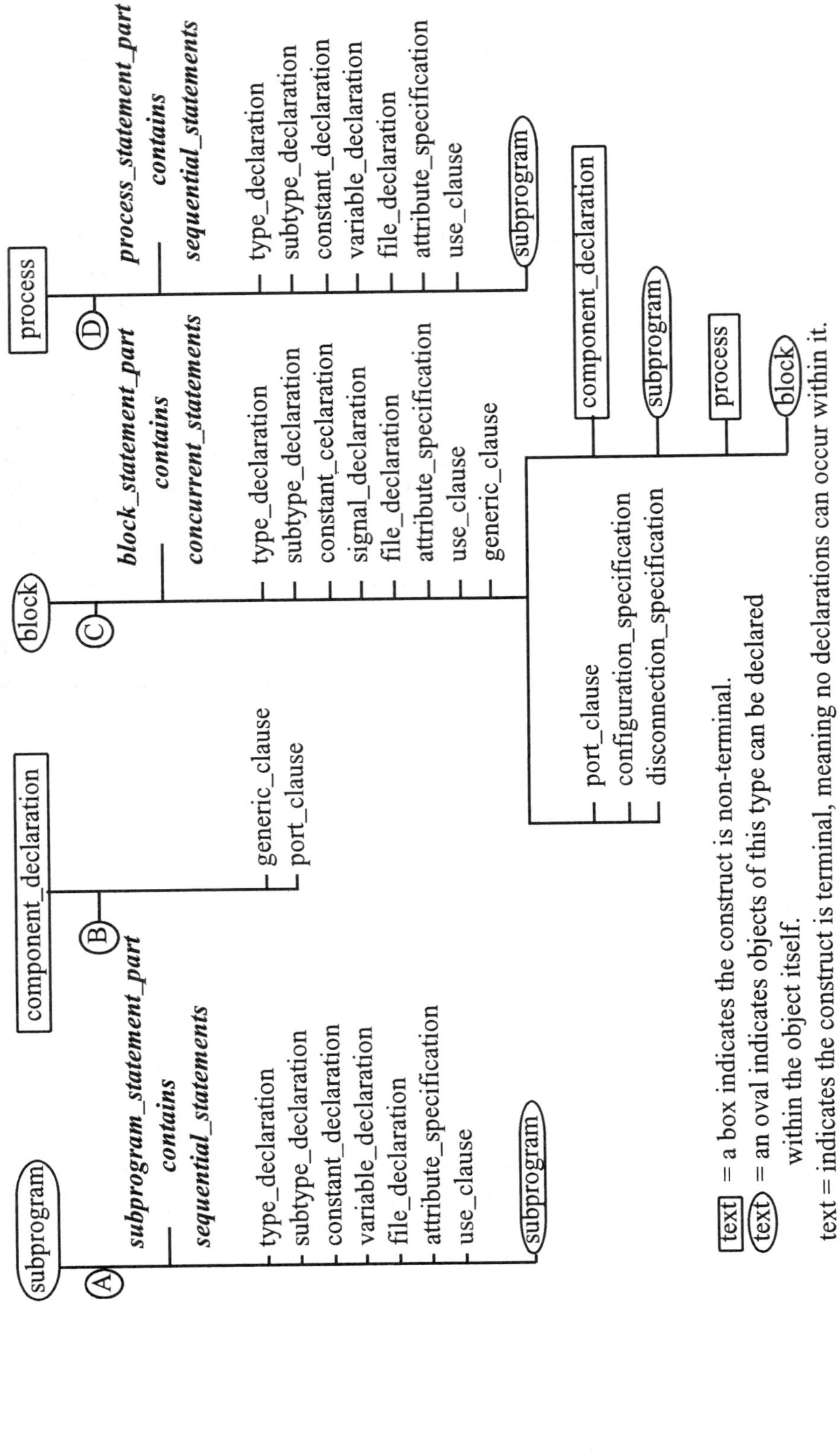
subprogram
A
subprogram_statement_part
contains
sequential_statements
type_declaration
subtype_declaration
constant_declaration
variable_declaration
file_declaration
attribute_specification
use_clause
subprogram
component_declaration
B
generic_clause
port_clause
block
C
block_statement_part
contains
concurrent_statements
type_declaration
subtype_declaration
constant_ceclaration
signal_declaration
file_declaration
attribute_specification
use_clause
generic_clause
port_clause
configuration_specification
disconnection_specification
component_declaration
subprogram
process
block
process
D
process_statement_part
contains
sequential_statements
type_declaration
subtype_declaration
constant_declaration
variable_declaration
file_declaration
attribute_specification
use_clause
subprogram
text = a box indicates the construct is non-terminal.
text = an oval indicates objects of this type can be declared within the object itself.
text = indicates the construct is terminal, meaning no declarations can occur within it.

Appendix C Language Changes (1993)

The working group within the IEEE that maintains VHDL standard is the VHDL Analysis and Standardisation Group (VASG). The VASG organised an international effort to re-standardise version VHDL-1987 and arrived at a revised standard in December 1992.

The changes were grouped into 5 topics - syntax and consistency, modelling, synthesis, packages and visibility, and open. The objective of the language design phase was to produce language change specifications (LCS) that satisfy the requirements for change. 48 LCSs were written and, after balloting, these were included, with a few exceptions, in ANSI/IEEE Std 1076-1993.

VHDL has been extended to provide greater consistency and integration capability. New capabilities in the 1993 version include groups (LCS-13), global or SHARED variables (LCS-46), hierarchical pathnames (LCS-03), and a facility to include foreign models (LCS-26) in a VHDL description. Some of the existing capabilities were extended or modified to facilitate initial and incremental creation of a design hierarchy. New shift/rotate operators were added to the language (See Section 3.1). The delay model of the language was modified to support pulse rejection. The syntactic consistency of the language was enhanced. Resolutions of ambiguities and inconsistencies addressed by the Issue Screening and Analysis Committee (ISAC) of the VASG were incorporated into the 1993 revision of the language.

The following is a representative, but not exhaustive, set of the changes to be found in VHDL-1993.

Section 2: The following new clauses have been added:

LCS-43: LRM 1993, 2.1.1.3, describing file parameters.

This LCS makes significant improvements to File I/O.

file becomes a fourth class of object, LRM 1993, **Section 4,** 4.3.

MODE is provided with new semantics.

OPEN and CLOSE are explicitly defined for files.

The concept of an **impure function** is introduced.

The introduction of files as a fourth class of object means that they are no longer treated as a variable with exceptions and provides a simple solution to the ENDFILE problem (See Section 15.2).

Explicit OPEN and CLOSE have been provided to giver the modeller more control over when the file must exist and be created. The implicit opening of the file in the file declaration is still allowed but it is now also possible to open the file in three modes controlled by a new enumerated type:

-- The predefined types for opening files:

type FILE_OPEN_KIND **is** (

```
READ_MODE,          -- Resulting access mode is read-only.
WRITE_MODE,         -- Resulting access mode is write-only.
APPEND_MODE) ;      -- Resulting access mode is write-only; information
                    -- is appended to the end of the existing file.

-- The predefined types for checking file open status:
type FILE_OPEN_STATUS is (
OPEN_OK,            -- File open was successful.
STATUS_ERROR,       -- File object was already open.
NAME_ERROR,         -- External file not found or inaccessible.
MODE_ERROR) ;       -- Could not open file with requested access
                    -- mode.
```

The extension does not cover random access files, for general portability reasons.

Finally, the concept of impure functions allows functions to be written which access global file objects.

LCS-22: LRM-1993, 2.3.2 : Adds the concept of Signatures.

A signature distinguishes between overloaded subprograms and overloaded enumeration literals based on their parameter and result type profiles. A signature can be used in an attribute name, entity designator, or alias declaration.

signature ::= [[type_mark { , type_mark}] [**return** type_mark]]

NB. The initial and terminal brackets are part of the syntax of signatures and do not indicate that the entire right-hand side of the production is optional (See Section 4.5.1).

Section 4: The following new clauses have been added:

LCS-13: LRM 1993 4.6 and 4.7, describing groups:

Groups extend the concept of attributes. They provide the ability to tag data to multiple objects rather than the single object for an attribute.

A group template defines the allowable classes of named entities that can appear in a group.

```
group_template_declaration ::=
    group identifier is (entity_class_entry_list) ;--Parentheses enclose
                                                    -- mandatory parts.
    entity_class_entry_list ::=
      entity_class_entry { , entity_class_entry} -- braces indicate that zero or
                              -- more of the contained parts may be used.
      entity_class_entry ::= entity_class [ <>] -- Square brackets indicate
                                                 -- optional parts.
```

A group declaration declares a group, a named collection of named entities.

group_declaration ::=

group identifier : *group_template*_name (group_constituent_list) ;
group_constituent_list ::= group_constituent { , group_constituent }
group_constituent ::= name | character_literal

LCS-46: LRM 1993, 4.4.1.3: Adds shared (global) variables:
variable declaration::=
[shared] variable identifier_list : subtype_indication [:= expression] ;

More than one process may access a shared variable. Every access to a shared variable must be made within a protected region which protects that variable.

Section 7 : New clauses describing the shift operators and the sign operators have been added: LCS-24: LRM 1993, 7.2.3: Provides 6 new shift operators for bit_vectors.

Operator	Operation	Left operand type	Right operand type	Result type
sll	Shift left logical	Any one_dimensional array type whose element type is BIT or BOOLEAN	INTEGER	Same as left
srl	Shift right logical	Any one_dimensional array type whose element type is BIT or BOOLEAN	INTEGER	Same as left
sla	Shift left arithmetic	Any one_dimensional array type whose element type is BIT or BOOLEAN	INTEGER	Same as left
sra	Shift right arithmetic	Any one_dimensional array type whose element type is BIT or BOOLEAN	INTEGER	Same as left
rol	Rotate left logical	Any one_dimensional array type whose element type is BIT or BOOLEAN	INTEGER	Same as left
ror	Rotate right logical	Any one_dimensional array type whose element type is BIT or BOOLEAN	INTEGER	Same as left

The shift operators are defined for any one-dimensional array type whose element type is either of the predefined types BIT or BOOLEAN.

LRM 1993, 7.2.5: Sign operators:

Takes the last paragraph and Note of LRM 1987, 7.2.3 and restates these as a new clause.

LCS-15: 7.4.2: Adds a modified definition of globally static primaries and clarifies issues relating to globally static expressions.

LCS-34: Makes a number of changes (clarifications) to locally and globally static expressions

Section 8 : Adds a new clause describing the report statement:

LCS-16: LRM 1993, 8.3 : Clarifies the **report** clause in relation to the Assertion statement. The report statement displays a message.

report_statement ::=
    [label :]
        report expression
            [**severity** expression] ;

The **report** statement expression must be of the predefined type STRING.

The new clause provides the ability to log assertion-like messages without the need for the assertion condition. This provides a facility for issuing messages from sequential statements.

Section 9 : Adds a new clause describing the instantiation of an entity:

Section 10 : LCS-01: LRM 1993, 10.1 : Adds a generate statement to the declarative region.

LRM 1993, 9.6.2. : Separates descriptions of component instantiations from those of entity instantiations.

Section 13 : Adds two new clauses describing basic and extended identifiers:

LCS-39: LRM 1993, 13.3.1 and 13.3.2 : A basic identifier consists only of letters, digits, and underlines.

basic_identifier ::=
    letter { [underline] letter_or_digit }

    letter_or_digit ::= letter | digit

    letter ::= upper_case_letter | lower_case_letter

All characters of a basic identifier are significant, including any underline character inserted between a letter or digit and an adjacent letter or digit.

Extended identifiers may contain any graphic character.

extended_identifier ::
    \graphic character { graphic_character }\

Example:

    \BUS\   \bus\   -- two different identifiers, neither of which is
                              -- the reserved word **bus**.

This addition to the VHDL enhances its use as a tool integration language by avoiding translation problems, and provides improved modelling capability for entity names staring with digits and signal names which are VHDL keywords (See

Exercise 5.6.7).

Section 14 : Adds a number of predefined attributes:

LCS-03: LRM 1993, 14.1 : Adds E'SIMPLE_NAME, E'INSTANCE_NAME and E'PATH_NAME as enhancements to the modelling capability. These attributes return, as Type String, the name of the object. This is useful when issuing messages from the model, such as setup/hold violations. They may also be important in providing hooks for back annotation.

LCS-14: improves the description of the attributes B'BEHAVIOR and B'STRUCTURE and adds:

T'ASCENDING: to the list of attributes for subtype T. Returns type Boolean.

T'IMAGE: Returns type String.

T'VALUE: Returns base type of T.

A'ASCENDING: to the list of attributes for array objects. Returns type Boolean.

S'DRIVING: to the list of attributes for static signal names. Returns type Boolean.

S'DRIVING_VALUE: Returns base type of S.

LCS-26: LRM 1993, 14.2 Package STANDARD : Adds the 'FOREIGN attribute:

attribute FOREIGN : STRING

The 'FOREIGN attribute may be associated only with architectures or with subprograms. In the latter case, the attribute specification must appear in the declarative part in which the subprogram is declared. The STRING value can be used to supply implementation specific information, for example, the name of the foreign subprogram.

Miscellaneous clause changes:

LCS-02: Which was to add a new attribute 'NET appears not to have been implemented although the word net is identified in the Glossary (LRM 1993 Annex B) as *a collection of drivers, signals (including ports and implicit signals), conversion functions, and resolution functions that connect different processes.* Initialisation of a net occurs after elaboration, and a net is updated during each simulation cycle.

LCS-04: LRM 1993, 9.2 : Adds the reserved word **postponed** to the process statement. The time at which a process executes after being resumed by a **wait** statement differs depending on whether the process is postponed or not postponed. When a non-postponed process is resumed, it executes in the current simulation cycle. When a postponed process is resumed, it does not execute until a simulation cycle occurs in which the next simulation is not a delta cycle. In this way, a postponed process accesses the values of signals that are the "final" values at the current simulation time. This solves the problem that assertion violations are often delta-cycle sensitive, where the intended semantics are required to check when values are stabilised, by postponing an assertion violation to a special last delta cycle.

LCS-06: LRM 1993, 14.1 :Provides user defined signal attributes. This enables the user to define signal attributes which detect, for example, rising edges on signals.

LCS-10: LRM 1993, 7.2 : Adds an **xnor** operator for consistency of logical operators. It is not completely backwards compatible as it introduces the new reserved word **xnor.**

LCS-11: LRM 1993, 4.3 : Eases the problem of associating composite types which differ only in the resolution of sub-elements. This frequently occurs with arrays of logic and arrays of resolved logic where VHDL requires the two array types to be different types. The problem is resolved by allowing type conversions on association lists. For a given object of a composite type, each subelement of that object is itself an object of the same class and subclass, if any, as the given object. The value of a composite object is the aggregation of the values of its subelements.

LCS-17: LRM 1993, 3.1.1.1 : Extends the predefined type CHARACTER from the 7-bit ISO 646 character set to the 8-bit 8859-1 character set. This is almost upwards compatible, although 'HIGH and 'RIGHT will be different. The extension is from the 128 ASCII set to 256 characters (LATIN-1).

LCS-18: LRM 1993, 3.2.1 : Resolves a problem which could prevent the concatenation of **downto** arrays. The statement:

variable A, B : bit_vector(7 **downto** 0) ;
B := A(6 **downto** 3) & A(3 **downto** 0) ;

gives a run-time error as the NATURAL index of BIT_VECTOR does not allow -1. The language change resolves this by making **downto** operations mirror the **to** operations.

LCS-19: LRM 1993, 8.4 : Adds the new reserved word **inertial** to provide an explicit delay paradigm which bridges the concepts of **transport** delay and the *implicit* inertial delay of 1076-1987. For example, the assignment:

output_sig <= **use** 5 ns **inertial** input_sig **after** 10 ns ;

specifies that the total propagation delay is 10 ns but that a pulse of less than 5 ns would be rejected. This provides an effective shorthand for specifying real circuit behaviour.

LCS-20: LRM 1993, 7.5 : removes the need for *exact* evaluation of *static* universal expressions, making them consistent with run-time universal expressions. This removes a great burden from implementations and brings the LRM into line with many existing implementations.

LCS-23: LRM 1993, 2.3 : Allows overloading of resolution functions; defining the rules used to select the appropriate resolution function.

A reference to an overloaded resolution function name in a subtype indication is ambiguous (and is therefore an error) if the name of the function, the number of formal parameters, the result type, and the relationships between the result type and the types of the formal parameters(as defined in 2.4) are not sufficient to identify exactly one (overloaded) subprogram specification.

LRM 1993, 2.4 : A resolution function must be a pure function; moreover, it must have a single input parameter of class **constant** that is a one-dimensional, unconstrained array whose element type is that of the resolved signal. The type of the return value of the function must also be that of the signal. Errors occur at the place of the subtype indication containing the resolution function name if any of these checks fail.

LCS-25: LRM 1993, 5.2.1, 9.6 : Allows direct instantiation of a component by using the entity name:

binding_indication ::=
  [**use** entity_aspect]
  [generic_aspect]
  [port_map_aspect]

This provides much improved usability, particularly in a top-down design environment. It saves replication of data where the component declaration is an almost exact copy of the entity declaration.

LCS-30: LRM 1993, 14.2 : Defines a new time sub-type in package STANDARD which is *non-negative*:

subtype DELAY_LENGTH **is** TIME **range** 0 fs **to** TIME'HIGH ;

for use as the return value of NOW and parameters for 'STABLE and 'DELAYED where it does not make sense to consider negative values:

impure function NOW **return** DELAY_LENGTH ;

LCS-34: LRM 1993, 7.4.1 : Makes a number of changes to locally and globally static expressions. These include:

A literal of type STANDARD.TIME is not allowed as a locally static primary as implementations tend to set a limit on TIME just prior to simulation. Hence the validity of a TIME quantity cannot be checked at analyse time.

LCS-35: LRM 1993, 4.3.2.1 : Clarifies the condition that the name of an interface object cannot be used as a constraint or default expression within that same interface list, such that, for example:

generic	*(G1: INTEGER ; G2: INTEGER := G1) ;*	*-- illegal*
port	*(P1: STRING ; P2: STRING(P1'RANGE) ;*	*-- illegal*
procedure	*(Y1,Y2: INTEGER; INTEGER **range** Y1 **to** Y2) ;*	*-- illegal*

LCS-36: LRM 1993, 4.3.2.2 : Forces the specification of a type conversion and an association to be consistent with the flow of data.

The formal part of a named element association may be in the form of a function call, where the single argument of the function is the formal designator itself, if and only if the mode of the formal is **out, inout, buffer,** or **linkage** and if the actual is not **open.**

LCS-42: LRM 1993, 1.1, 4.5 : Provides for uniform construct bracketing in the language. The main problem was that, following the reserved word **end,** the language was not consistent as to whether a keyword denoting the entity type, or an identifier denoting the entity name was required. For example:

```
entity E1...
end E1 ;
component C1...
end component ;
```

This is solved by adding the optional *reserved_word* to those endings which did not allow entity type *reserved_words,* and the optional identifier to those endings which did not allow entity identifiers.

```
entity_declaration ::=
    entity identifier is
            entity_header
            entity_declarative_part
    [begin
            entity_statement_part ]
    end [entity] [ entity_simple_name ] ;
component_declaration ::=
    component identifier is
            [ local_generic_clause ]
            [ local_port_clause ]
    end component [ component_simple_name ] ;
```

If a simple name appears at the end of an entity/component declaration, it must repeat the identifier of the entity/component declaration.

LCS-45: LRM 1993, 5.2.2 : Clarifies default binding without configuration declarations. Many interpreted the 1987 language as requiring a configuration declaration, but thought it an unnecessary restriction. Visibility of the entity aspect created a problem as the component declaration would often hide the entity declaration. The binding rule has now been enhanced and states:

A *visible entity declaration* is either

a) An entity declaration that has the same simple name as that of the instantiated component and that is directly visible, or

b) An entity declaration that has the same simple name as that of the instantiated component and that would be directly visible in the absence of a directly visible component declaration with the same simple name as that of the entity declaration.

Index